Cram101 Textbook Outlines to accompany:

Geology In The Field

Compton, 1st Edition

An Academic Internet Publishers (AIPI) publication (c) 2007.

You have a discounted membership at www.Cram101.com with this book.

Get all of the practice tests for the chapters of this textbook, and access in-depth reference material for writing essays and papers. Here is an example from a Cram101 Biology text:

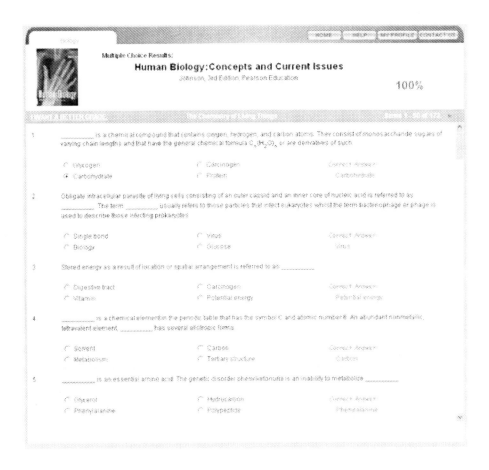

When you need problem solving help with math, stats, and other disciplines, www.Cram101.com will walk through the formulas and solutions step by step.

With Cram101.com online, you also have access to extensive reference material.

You will nail those essays and papers. Here is an example from a Cram101 Biology text:

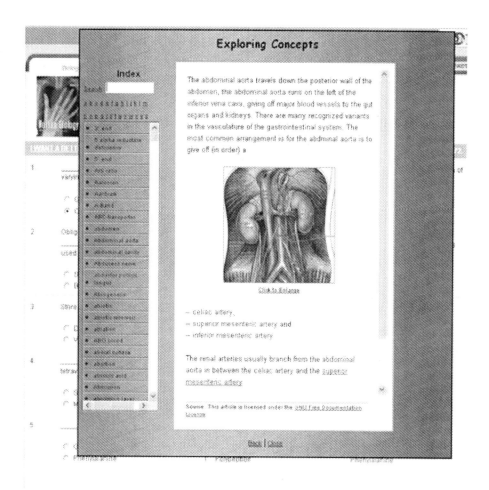

Visit **www.Cram101.com**, click Sign Up at the top of the screen, and enter DK73DW725 in the promo code box on the registration screen. Access to www.Cram101.com is normally $9.95, but because you have purchased this book, your access fee is only $4.95. Sign up and stop highlighting textbooks forever.

Learning System

Cram101 Textbook Outlines is a learning system. The notes in this book are the highlights of your textbook, you will never have to highlight a book again.

How to use this book. Take this book to class, it is your notebook for the lecture. The notes and highlights on the left hand side of the pages follow the outline and order of the textbook. All you have to do is follow along while your intructor presents the lecture. Circle the items emphasized in class and add other important information on the right side. With Cram101 Textbook Outlines you'll spend less time writing and more time listening. Learning becomes more efficient.

Cram101.com Online

Increase your studying efficiency by using Cram101.com's practice tests and online reference material. It is the perfect complement to Cram101 Textbook Outlines. Use self-teaching matching tests or simulate in-class testing with comprehensive multiple choice tests, or simply use Cram's true and false tests for quick review. Cram101.com even allows you to enter your in-class notes for an integrated studying format combining the textbook notes with your class notes.

Visit **www.Cram101.com**, click Sign Up at the top of the screen, and enter **DK73DW725** in the promo code box on the registration screen. Access to www.Cram101.com is normally $9.95, but because you have purchased this book, your access fee is only $4.95. Sign up and stop highlighting textbooks forever.

Geology In The Field
Compton, 1st

CONTENTS

Philosophy	Philosophy is the discipline concerned with the questions of how one should live ; what sorts of things exist and what are their essential natures ; what counts as genuine knowledge; and what are the correct principles of reasoning.
Geology	Geology is the science and study of the solid matter that constitute the Earth. Encompassing such things as rocks, soil, and gemstones, geology studies the composition, structure, physical properties, history, and the processes that shape Earth's components.
Volcano	A volcano is an opening, or rupture, in the Earth's surface or crust, which allows hot, molten rock, ash and gases to escape from deep below the surface.
Lava	Lava is molten rock expelled by a volcano during an eruption. When first extruded from a volcanic vent, it is a liquid at temperatures from 700 °C to 1,200 °C.
Intrusion	An intrusion is a body of igneous rock that has crystallized from a molten magma below the surface of the Earth.
Metamorphism	Metamorphism can be defined as the solid state recrystallisation of pre-existing rocks due to changes in heat and/or pressure and/or introduction of fluids. There will be mineralogical, chemical and crystallographic changes. Metamorphism produced with increasing pressure and temperature conditions is known as prograde metamorphism. Conversely, decreasing temperatures and pressure characterize retrograde metamorphism.
Mineral	A mineral is a naturally occurring substance formed through geological processes that has a characteristic chemical composition, a highly ordered atomic structure and specific physical properties. A rock, by comparison, is an aggregate of minerals and need not have a specific chemical composition. Minerals range in composition from pure elements and simple salts to very complex silicates with thousands of known forms.
Igneous	Igneous rocks form when molten rock, magma, cools and solidifies, with or without crystallization, either below the surface as intrusive, plutonic rocks or on the surface as extrusive, volcanic, rocks.
Matter	Matter is the substance of which physical objects are composed. Matter can be solid, liquid, plasma or gas. It constitutes the observable universe.
Geological Society of America	The Geological Society of America is a nonprofit organization dedicated to the advancement of the geosciences. The society was founded in New York in 1888 by James Hall, James D. Dana, and Alexander Winchell, and has been headquartered at 3300 Penrose Place, Boulder, Colorado since 1968. As of 2007, the society has over 21,000 members in more than 85 countries. The stated mission is "to advance the geosciences, to enhance the professional growth of its members, and to promote the geosciences in the service of humankind".
Earth science	Earth science is an all-embracing term for the sciences related to the planet Earth. It is arguably a special case in planetary science, being the only known life-bearing planet. There are both reductionist and holistic approaches to Earth science. The major historic disciplines use physics, geology, geography, mathematics, chemistry, and biology to build a quantitative understanding of the principal areas or spheres of the Earth system.
Geologic map	A geologic map is a special-purpose map made to show geological features. The stratigraphic contour lines are drawn on the surface of a selected deep stratum, so that they can show the topographic trends of the strata under the ground. It is not always possible to properly show this when the strata are extremely fractured, mixed, in some discontinuities, or where they are otherwise disturbed.
Rapid	A rapid is a section of a river of relatively steep gradient causing an increase in water flow and turbulence. A rapid is a hydrological feature between a run and a cascade. It is characterized by the river becoming shallower and having some rocks exposed above the flow surface.
Fossil	Fossils are the mineralized or otherwise preserved remains or traces of animals, plants, and other organisms. The totality of fossils, both discovered and undiscovered, and their placement in

fossiliferous rock formations and sedimentary layers is known as the fossil record.

Foliated metamorphic rock

Foliated metamorphic rock has penetrative planar fabric present within it. It is common to rocks affected by regional metamorphic compression typical of orogenic belts.

Climate

Climate is the average and variations of weather over long periods of time. Climate zones can be defined using parameters such as temperature and rainfall.

Hardness

Hardness is the characteristic of a solid material expressing its resistance to permanent deformation.

4

Go to **Cram101.com** for the Practice Tests for this Chapter.

Mineral	A mineral is a naturally occurring substance formed through geological processes that has a characteristic chemical composition, a highly ordered atomic structure and specific physical properties. A rock, by comparison, is an aggregate of minerals and need not have a specific chemical composition. Minerals range in composition from pure elements and simple salts to very complex silicates with thousands of known forms.
Luster	Luster is a description of the way light interacts with the surface of a crystal, rock, or mineral. For example, a diamond is said to have an adamantine luster and pyrite is said to have a metallic luster.
Crystal	A crystal is a solid in which the constituent atoms, molecules, or ions are packed in a regularly ordered, repeating pattern extending in all three spatial dimensions. Most metals encountered in everyday life are polycrystals. Crystals are often symmetrically intergrown to form crystal twins.
Cleavage	Cleavage, in mineralogy, is the tendency of crystalline materials to split along definite planes, creating smooth surfaces.
Rotation	A rotation is a movement of an object in a circular motion. A two-dimensional object rotates around a center of rotation. A three-dimensional object rotates around a line called an axis. A circular motion about an external point, e.g. the Earth about the Sun, is called an orbit or more properly an orbital revolution.
Coal	Coal is a fossil fuel formed in swamp ecosystems where plant remains were saved by water and mud from oxidization and biodegradation. It is a sedimentary rock, but the harder forms, such as anthracite coal, can be regarded as metamorphic rocks because of later exposure to elevated temperature and pressure. It is composed primarily of carbon along with assorted other elements, including sulfur.
Potassium	Potassium is a chemical element. It is a soft silvery-white metallic alkali metal that occurs naturally bound to other elements in seawater and many minerals. It oxidizes rapidly in air and is very reactive, especially towards water. In many respects, it and sodium are chemically similar, although organisms in general, and animal cells in particular, treat them very differently.
Feldspar	Feldspar is the name of a group of rock-forming minerals which make up as much as sixty percent of the Earth's crust. Feldspars crystallize from magma in both intrusive and extrusive rocks, and they can also occur as compact minerals, as veins, and are also present in many types of metamorphic rock.
Landfill	A landfill, is a site for the disposal of waste materials by burial and is the oldest form of waste treatment.
Contour line	A contour line shows elevation. A contour line for a function of two variables is a curve connecting points where the function has a same particular value. A contour map is a map illustrated with contour lines, for example a topographic map.
Rapid	A rapid is a section of a river of relatively steep gradient causing an increase in water flow and turbulence. A rapid is a hydrological feature between a run and a cascade. It is characterized by the river becoming shallower and having some rocks exposed above the flow surface.
Stratigraphy	Stratigraphy, a branch of geology, studies rock layers and layering. It is primarily used in the study of sedimentary and layered volcanic rocks. Stratigraphy includes two related subfields: lithologic or lithostratigraphy and biologic stratigraphy or biostratigraphy.

Go to **Cram101.com** for the Practice Tests for this Chapter.
And, **NEVER** highlight a book again!

Carbonate	In organic chemistry, a carbonate is a salt of carbonic acid.
Mineral	A mineral is a naturally occurring substance formed through geological processes that has a characteristic chemical composition, a highly ordered atomic structure and specific physical properties. A rock, by comparison, is an aggregate of minerals and need not have a specific chemical composition. Minerals range in composition from pure elements and simple salts to very complex silicates with thousands of known forms.
Alcatraz Island	Alcatraz Island is a small island located in the middle of San Francisco Bay in California, United States. It served as a lighthouse, then a military fortification, then a military prison followed by a federal prison until 1963, when it became a national recreation area.
Currents	Ocean currents are any more or less continuous, directed movement of ocean water that flows in one of the Earth's oceans.They are rivers of hot or cold water within the ocean. They are generated from the forces acting upon the water like the earth's rotation, the wind, the temperature and salinity differences and the gravitation of the moon.
Shale	Shale is a fine-grained sedimentary rock whose original constituents were clays or muds. It is characterized by thin laminae breaking with an irregular curving fracture, often splintery and usually parallel to the often-indistinguishable bedding plane.
Sandstone	Sandstone is a sedimentary rock composed mainly of sand-size mineral or rock grains. Most sandstone is composed of quartz and/or feldspar because these are the most common minerals in the Earth's crust. Like sand, sandstone may be any color, but the most common colors are tan, brown, yellow, red, gray and white.
Granite	Granite is a common and widely occurring type of intrusive, felsic, igneous rock. Granites are usually medium to coarsely crystalline, occasionally with some individual crystals larger than the groundmass forming a rock known as porphyry. Granites can be pink to dark gray or even black, depending on their chemistry and mineralogy.
Biotite	Biotite is a common phyllosilicate mineral within the mica group. Primarily a solid-solution series between the iron-endmember annite, and the magnesium-endmember phlogopite; more aluminous endmembers include siderophyllite.
Dolomite	Dolomite is the name of a sedimentary carbonate rock and a mineral, both composed of calcium magnesium carbonate found in crystals. Dolomite rock is composed predominantly of the mineral dolomite. Limestone that is partially replaced by dolomite is referred to as dolomitic limestone.
Scientific method	Scientific method is a body of techniques for investigating phenomena and acquiring new knowledge, as well as for correcting and integrating previous knowledge. It is based on gathering observable, empirical and measurable evidence subject to specific principles of reasoning,
Intrusion	An intrusion is a body of igneous rock that has crystallized from a molten magma below the surface of the Earth.
Sill	In geology, a sill is a tabular pluton that has intruded between older layers of sedimentary rock, beds of volcanic lava or tuff, or even along the direction of foliation in metamorphic rock. The term sill is synonymous with concordant intrusive sheet. This means that the sill does not cut across preexisting rocks. Contrast this with dikes.
Era	An era is a long period of time with different technical and colloquial meanings, and usages in language. It begins with some beginning event known as an epoch, epochal date, epochal event or epochal moment.
Fossil	Fossils are the mineralized or otherwise preserved remains or traces of animals, plants, and

Go to **Cram101.com** for the Practice Tests for this Chapter.

other organisms. The totality of fossils, both discovered and undiscovered, and their placement in fossiliferous rock formations and sedimentary layers is known as the fossil record.

Unconformity	An unconformity is a buried erosion surface separating two rock masses or strata of different ages, indicating that sediment deposition was not continuous. In general, the older layer was exposed to erosion for an interval of time before deposition of the younger, but the term is used to describe any break in the sedimentary geologic record.
Concretion	A concretion is a volume of sedimentary rock in which a mineral cement fills the porosity. They are often ovoid or spherical in shape, although irregular shapes also occur. They form within layers of sedimentary strata that have already been deposited. They usually form early in the burial history of the sediment, before the rest of the sediment is hardened into rock.
Hydrocarbon	In organic chemistry, a hydrocarbon is an organic compound consisting entirely of hydrogen and carbon. With relation to chemical terminology, aromatic hydrocarbons or arenes, alkanes, alkenes and alkyne-based compounds composed entirely of carbon or hydrogen are referred to as "Pure" hydrocarbons, whereas other hydrocarbons with bonded compounds or impurities of sulphur or nitrogen, are referred to as "impure", and remain somewhat erroneously referred to as hydrocarbons.
Cleavage	Cleavage, in mineralogy, is the tendency of crystalline materials to split along definite planes, creating smooth surfaces.
Blocks	Blocks in meteorology are large scale patterns in the atmospheric pressure field that are nearly stationary, effectively "blocking" or redirecting migratory cyclones. These blocks can remain in place for several days or even weeks, causing the areas affected by them to have the same kind of weather for an extended period of time.
Fault	Faults are planar rock fractures, which show evidence of relative movement. Large faults within the Earth's crust are the result of shear motion and active fault zones are the causal locations of most earthquakes. Earthquakes are caused by energy release during rapid slippage along faults. The largest examples are at tectonic plate boundaries but many faults occur far from active plate boundaries. Since faults do not usually consist of a single, clean fracture, the term fault zone is used when referring to the zone of complex deformation that is associated with the fault plane.
Foliated metamorphic rock	Foliated metamorphic rock has penetrative planar fabric present within it. It is common to rocks affected by regional metamorphic compression typical of orogenic belts.
Fold	The term fold is used in geology when one or a stack of originally flat and planar surfaces, such as sedimentary strata, are bent or curved as a result of plastic, i.e. permanent, deformation.
Paleontology	Paleontology, palaeontology or palæontology is the study of prehistoric life forms on Earth through the examination of plant and animal fossils. This includes the study of body fossils, tracks, burrows, cast-off parts, fossilised faeces, palynomorphs and chemical residues. See also paleoanthropology.
Trace fossil	A trace fossil is a structure preserved in sedimentary rocks that record biological activity. While we are most familiar with relatively spectacular, fossilized hard-part remains such as shells and bones, a trace fossil is often less dramatic, but nonetheless very important.
Invertebrate	Invertebrate is an English word that describes any animal without a spinal column.
Calcareous	Calcareous refers to a sediment, sedimentary rock, or soil type which is formed from or contains a high proportion of calcium carbonate in the form of calcite or aragonite.

Limestone	Limestone is a sedimentary rock composed largely of the mineral calcite. Limestone often contains variable amounts of silica in the form of chert or flint, as well as varying amounts of clay, silt and sand as disseminations, nodules, or layers within the rock. The primary source of the calcite in limestone is most commonly marine organisms. These organisms secrete shells that settle out of the water column and are deposited on ocean floors as pelagic ooze or alternatively is conglomerated in a coral reef.
Compaction	Compaction is the process of a material being more closely packed together.
Nodule	A nodule in petrology or mineralogy is an irregular rounded to spherical concretion. They are typically solid replacement bodies of chert or iron oxides formed during diagenesis of a sedimentary rock.
Diagenesis	In geology and oceanography, diagenesis is any chemical, physical, or biological change undergone by a sediment after its initial deposition and during and after its lithification, exclusive of surface alteration, weathering and metamorphism. These changes happen at relatively low temperatures and pressures and result in changes to the rock face=symbol>¢s original mineralogy and texture
Calcite	The carbonate mineral Calcite is a chemical or biochemical calcium carbonate and is one of the most widely distributed minerals on the Earth's surface. It is a common constituent of sedimentary rocks, limestone in particular. It is also the primary mineral in metamorphic marble
Weathering	Weathering is the process of breaking down rocks, soils and their minerals through direct contact with the atmosphere. Weathering occurs without movement. Two main classifications of weathering processes exist. Mechanical or physical weathering involves the breakdown of rocks and soils through direct contact with atmospheric conditions. The second classification, chemical weathering, involves the direct effect of atmospheric chemicals in the breakdown of rocks, soils and minerals.
Siltstone	Siltstone is a sedimentary rock which has a composition intermediate in grain size between the coarser sandstones and the finer mudstones and shales.
Mold	Mold refers to all species of microscopic fungi that grow in the form of multicellular filaments, called hyphae.
Mudstone	Mudstone is a fine-grained sedimentary rock whose original constituents were clays or muds. Grain size is up to 0.0625 mm with individual grains too small to be distinguished without a microscope.
Brachiopods	Brachiopods are a phylum of animals. They are sessile, two-shelled, marine animals with an external morphology resembling pelecypod mollusks of phylum Mollusca to which they are not closely related.
Chert	Chert is a fine-grained silica-rich cryptocrystalline sedimentary rock that may contain small fossils. It varies greatly in color from white to black, but most often manifests as gray, brown, grayish brown and light green to rusty red; its color is an expression of trace elements present in the rock, and both red and green are most often related to traces of iron.
Conglomerate	A conglomerate is a rock consisting of individual stones that have become cemented together. Conglomerates are sedimentary rocks consisting of rounded fragments and are thus differentiated from breccias, which consist of angular clasts. Both conglomerates and breccias are characterized by clasts larger than sand.
Gastropods	The gastropods are the largest and most successful class of mollusks. This class contains a vast number of marine and freshwater species as well as many terrestrial ones. Species include not only the snails and slugs, but also abalone, limpets, cowries, conch and most of

the other animals that produce seashells.

Cephalopods	The cephalopods are the mollusc class Cephalopoda characterized by bilateral body symmetry, a prominent head, and a modification of the mollusk foot, a muscular hydrostat, into the form of arms or tentacles. The class contains two extant subclasses. In the Coleoidea, the mollusk shell has been internalized or is absent; this subclass includes the octopuses, squid, and cuttlefish. In the Nautiloidea the shell remains; this subclass includes the nautilus.
Coral	Coral refer to marine animals from the class Anthozoa and exist as small sea anemone-like polyps, typically in colonies of many identical individuals. The group includes the important reef builders that are found in tropical oceans, which secrete calcium carbonate to form a hard skeleton.
Deposition	Deposition is the geological process whereby material is added to a landform. This is the process by which wind and water create a sediment deposit, through the laying down of granular material that has been eroded and transported from another geographical location.
Stratigraphy	Stratigraphy, a branch of geology, studies rock layers and layering. It is primarily used in the study of sedimentary and layered volcanic rocks. Stratigraphy includes two related subfields: lithologic or lithostratigraphy and biologic stratigraphy or biostratigraphy.
Pelagic	The pelagic is the part of the open sea or ocean that is not near the coast or sea floor. In contrast, the demersal zone comprises the water that is near to, and is significantly affected by, the coast or the sea floor.
Biozone	Biozone refers to intervals of geological strata that are defined on the basis of their characteristic fossil taxa.
Electron correlation	Electron correlation refers to the interaction between electrons in a quantum system whose electronic structure is being considered.
Delta	A delta is a landform where the mouth of a river flows into an ocean, sea, desert, estuary or lake. It builds up sediment outwards into the flat area which the river face=symbol>¢s flow encounters transported by the water and set down as the currents slow.
Lake	A lake is a body of water or other liquid of considerable size contained on a body of land. A vast majority are fresh water, and lie in the Northern Hemisphere at higher latitudes. Most have a natural outflow in the form of a river or stream, but some do not, and lose water solely by evaporation and/or underground seepage.
Coal	Coal is a fossil fuel formed in swamp ecosystems where plant remains were saved by water and mud from oxidization and biodegradation. It is a sedimentary rock, but the harder forms, such as anthracite coal, can be regarded as metamorphic rocks because of later exposure to elevated temperature and pressure. It is composed primarily of carbon along with assorted other elements, including sulfur.
Claystone	Claystone is a geological term used to describe a sedimentary rock that is composed primarily of clay-sized particles.
Conodontophora	The conodontophora is an extinct worm-like animal with distinctive conical or multi-denticulate teeth made of apatite. The teeth of these animals, is almost always the only parts preserved as fossils. The teeth show complex specialized structures, and survived through the ages and the fossilization process due to their resilient phosphatic chemical composition; the teeth were probably used to filter out plankton and pass it down the throat. Conodontophora and their presumed relatives are known from the Cambrian to the Late Triassic.

Sediment	Sediment is any particulate matter that can be transported by fluid flow and which eventually is deposited as a layer of solid particles on the bed or bottom of a body of water or other liquid.
Clastic	Clastic rocks are rocks formed from fragments of pre-existing rock.
Shale	Shale is a fine-grained sedimentary rock whose original constituents were clays or muds. It is characterized by thin laminae breaking with an irregular curving fracture, often splintery and usually parallel to the often-indistinguishable bedding plane.
Silt	Silt is soil or rock derived granular material of a specific grain size. Silt may occur as a soil or alternatively as suspended sediment in a water column of any surface water body. It may also exist as deposition soil at the bottom of a water body.
Quartz	Quartz is the second most common mineral in the Earth's continental crust. It is made up of a lattice of silica tetrahedra. Quartz belongs to the rhombohedral crystal system. In nature quartz crystals are often twinned, distorted, or so intergrown with adjacent crystals of quartz or other minerals as to only show part of this shape, or to lack obvious crystal faces altogether and appear massive.
Feldspar	Feldspar is the name of a group of rock-forming minerals which make up as much as sixty percent of the Earth's crust. Feldspars crystallize from magma in both intrusive and extrusive rocks, and they can also occur as compact minerals, as veins, and are also present in many types of metamorphic rock.
Particle size	Particle size refers to the diameter of individual grains of sediment, or the lithified particles in clastic rocks. The term may also be applied to other granular materials. This is different from the crystallite size, which is the size of a single crystal inside the particles or grains.
Debris flow	Debris flow often refers to mudslides, mudflows, jökulhlaups, or debris avalanches. They consist primarily of geological material mixed with water. They may be generated when hillside colluvium or landslide material becomes rapidly saturated with water and flows into a channel.
Abrasion	Abrasion is mechanical scraping of a rock surface by friction between rocks and moving particles during their transport in wind, glacier, waves, gravity or running water.
Mica	The mica group of sheet silicate minerals includes several closely related materials having highly perfect basal cleavage. All are monoclinic with a tendency towards pseudo-hexagonal crystals and are similar in chemical composition. The highly perfect cleavage, which is the most prominent characteristic of mica, is explained by the hexagonal sheet-like arrangement of its atoms.
Fossil	Fossils are the mineralized or otherwise preserved remains or traces of animals, plants, and other organisms. The totality of fossils, both discovered and undiscovered, and their placement in fossiliferous rock formations and sedimentary layers is known as the fossil record.
Sedimentary	Sedimentary rock is one of the three main rock groups. Rock formed from these covers 75% of the Earth's land area, and includes common types such as chalk, limestone, dolomite, sandstone, and shale.
Currents	Ocean currents are any more or less continuous, directed movement of ocean water that flows in one of the Earth's oceans. They are rivers of hot or cold water within the ocean. They are generated from the forces acting upon the water like the earth's rotation, the wind, the temperature and salinity differences and the gravitation of the moon.

Go to **Cram101.com** for the Practice Tests for this Chapter.
And, **NEVER** highlight a book again!

Paleontologists	Paleontologists are people who study prehistoric life forms on Earth through the examination of plant and animal fossils. This includes the study of body fossils, tracks, burrows, cast-off parts, fossilised faeces, palynomorphs and chemical residues.
Coral	Coral refer to marine animals from the class Anthozoa and exist as small sea anemone-like polyps, typically in colonies of many identical individuals. The group includes the important reef builders that are found in tropical oceans, which secrete calcium carbonate to form a hard skeleton.
Mollusks	The mollusks are members of the large and diverse phylum, which includes a variety of familiar animals well-known for their decorative shells or as seafood. These range from tiny snails, clams, and abalone to larger organisms such as squid, cuttlefish and the octopus
Predator	A predator is an organism that feeds on another living organism or organisms known as prey. A predator may or may not kill their prey prior to or during the act of feeding on them.
Compaction	Compaction is the process of a material being more closely packed together.
Recrystalliz-tion	Recrystallization is an essentially physical process that has meanings in chemistry, metallurgy and geology. In geology, solid-state recrystallization is a metamorphic process that occurs under situations of intense temperature and pressure where grains, atoms or molecules of a rock or mineral are packed closer together, creating a new crystal structure.
Polymorphism	In biology, polymorphism can be defined as discontinuous variation in a single population, in other words, the occurrence of more than one form or type of individual (whether the differences are visible or solely biochemical). The most obvious example of polymorphism is the sexual dimorphism of most higher organisms.
Cementation	Cementation is the process of deposition of dissolved mineral components in the interstices of sediments. It is an important factor in the consolidation of coarse-grained clastic sedimentary rocks such as sandstones, conglomerates, or breccias during diagenesis or lithification. Cementing materials may include silica, carbonates, iron oxides, or clay minerals.
Carbonate	In organic chemistry, a carbonate is a salt of carbonic acid.
Oxide	An oxide is a chemical compound containing an oxygen atom and other elements. Most of the earth's crust consists of them. They result when elements are oxidized by air.
Iron	Iron is a chemical element metal. It is a lustrous, silvery soft metal. It and nickel are notable for being the final elements produced by stellar nucleosynthesis, and thus are the heaviest elements which do not require a supernova or similarly cataclysmic event for formation.
Cleavage	Cleavage, in mineralogy, is the tendency of crystalline materials to split along definite planes, creating smooth surfaces.
Diagenesis	In geology and oceanography, diagenesis is any chemical, physical, or biological change undergone by a sediment after its initial deposition and during and after its lithification, exclusive of surface alteration, weathering and metamorphism. These changes happen at relatively low temperatures and pressures and result in changes to the rock face=symbol>¢s original mineralogy and texture
Luster	Luster is a description of the way light interacts with the surface of a crystal, rock, or mineral. For example, a diamond is said to have an adamantine luster and pyrite is said to have a metallic luster.
Chert	Chert is a fine-grained silica-rich cryptocrystalline sedimentary rock that may contain small fossils. It varies greatly in color from white to black, but most often manifests as gray,

Go to **Cram101.com** for the Practice Tests for this Chapter.

brown, grayish brown and light green to rusty red; its color is an expression of trace elements present in the rock, and both red and green are most often related to traces of iron.

Diatoms

Diatoms are a major group of eukaryotic algae, and are one of the most common types of phytoplankton. Most diatoms are unicellular, although some form chains or simple colonies. A characteristic feature of diatom cells is that they are encased within a unique cell wall made of silica called a frustule.

Radiolarian

Radiolarian refers to an amoeboid protozoa that produce intricate mineral skeletons, typically with a central capsule dividing the cell into inner and outer portions, called endoplasm and ectoplasm. They are found as zooplankton throughout the ocean, and because of their rapid turn-over of species, their tests are important diagnostic fossils found from the Cambrian onwards.

Amorphous

An amorphous solid is a solid in which there is no long-range order of the positions of the atoms. These materials are often prepared by rapidly cooling molten material, such as glass. The cooling reduces the mobility of the material's molecules before they can pack into a more thermodynamically favorable crystalline state.

Diatomaceous earth

Diatomaceous earth is a naturally occurring, soft, chalk-like sedimentary rock that is easily crumbled into a fine white to off-white powder. This powder has an abrasive feel, similar to pumice powder, and is very light, due to its high porosity.

Impurities

Impurities are substances inside a confined amount of liquid, gas, or solid, which differ from the chemical composition of the material or compound.

Mineral

A mineral is a naturally occurring substance formed through geological processes that has a characteristic chemical composition, a highly ordered atomic structure and specific physical properties. A rock, by comparison, is an aggregate of minerals and need not have a specific chemical composition. Minerals range in composition from pure elements and simple salts to very complex silicates with thousands of known forms.

Calcite

The carbonate mineral Calcite is a chemical or biochemical calcium carbonate and is one of the most widely distributed minerals on the Earth's surface. It is a common constituent of sedimentary rocks, limestone in particular. It is also the primary mineral in metamorphic marble

Sulfide

The term sulfide refers to several types of chemical compounds containing sulfur in its lowest oxidation number of −2.

Dolomite

Dolomite is the name of a sedimentary carbonate rock and a mineral, both composed of calcium magnesium carbonate found in crystals. Dolomite rock is composed predominantly of the mineral dolomite. Limestone that is partially replaced by dolomite is referred to as dolomitic limestone.

Apatite

Apatite is a group of phosphate minerals, usually referring to hydroxylapatite, fluorapatite, and chlorapatite, named for high concentrations of OH-, F-, or Cl- ions, respectively, in the crystal.

Aragonite

Aragonite is a carbonate mineral. It and the mineral calcite are the two common, naturally occurring polymorphs of calcium carbonate. The crystal lattice of aragonite differs from that of calcite, resulting in a different crystal shape, an orthorhombic system with acicular crystals. By repeated twinning pseudo-hexagonal forms result. It may be columnar or fibrous, occasionally in branching stalactitic forms called flos-ferri from their association with the ores at the Carthinian iron mines.

Oolite

An oolite is a sedimentary rock formed from ooids, spherical grains composed of concentric layers.

Conglomerate	A conglomerate is a rock consisting of individual stones that have become cemented together. Conglomerates are sedimentary rocks consisting of rounded fragements and are thus differentiated from breccias, which consist of angular clasts. Both conglomerates and breccias are characterized by clasts larger than sand.
Breccia	Breccia is a rock composed of angular fragments of rocks or minerals in a matrix, that is a cementing material, that may be similar or different in composition to the fragments.
Clasts	Clastic sedimentary rocks are rocks composed predominantly of broken pieces or clasts of older weathered and eroded rocks.
Alcatraz Island	Alcatraz Island is a small island located in the middle of San Francisco Bay in California, United States. It served as a lighthouse, then a military fortification, then a military prison followed by a federal prison until 1963, when it became a national recreation area.
Gravel	Gravel is rock that is of a certain particle size range. In geology, gravel is any loose rock that is at least two millimeters in its largest dimension and no more than 75 millimeters.
Siltstone	Siltstone is a sedimentary rock which has a composition intermediate in grain size between the coarser sandstones and the finer mudstones and shales.
Clay	Clay is a term used to describe a group of hydrous aluminium phyllosilicate minerals, that are typically less than 2 micrometres in diameter. Clay consists of a variety of phyllosilicate minerals rich in silicon and aluminium oxides and hydroxides which include variable amounts of structural water. Clays are generally formed by the chemical weathering of silicate-bearing rocks by carbonic acid but some are formed by hydrothermal activity.
Claystone	Claystone is a geological term used to describe a sedimentary rock that is composed primarily of clay-sized particles.
Ooid	Ooid refers to small spheroidal "coated" grains, usually composed of calcium carbonate, but sometimes made up of iron or phosphate minerals. They usually form on the sea floor, most commonly in shallow tropical seas. After being buried under additional sediment, these ooid grains can be cemented together to form a sedimentary rock called an oolite.
Limestone	Limestone is a sedimentary rock composed largely of the mineral calcite. Limestone often contains variable amounts of silica in the form of chert or flint, as well as varying amounts of clay, silt and sand as disseminations, nodules, or layers within the rock. The primary source of the calcite in limestone is most commonly marine organisms. These organisms secrete shells that settle out of the water column and are deposited on ocean floors as pelagic ooze or alternatively is conglomerated in a coral reef.
Algae	Algae encompass several groups of relatively simple living aquatic organisms that capture light energy through photosynthesis, using it to convert inorganic substances into organic matter.
Dolostone	Dolostone is a sedimentary carbonate rock that contains a high percentage of the mineral dolomite. It is usually referred to as dolomite rock. In old U.S.G.S. publications it was referred to as magnesian limestone.
Nodule	A nodule in petrology or mineralogy is an irregular rounded to spherical concretion. They are typically solid replacement bodies of chert or iron oxides formed during diagenesis of a sedimentary rock.
Coprolite	Coprolite is the name given to the mineral that results when human or animal dung is fossilized. Coprolite may range in size from the size of a BB all the way up to that of a large appliance.
Cryptocrysta-line	Cryptocrystalline is a rock texture which is so finely crystalline, that is, made up of such minute crystals that its crystalline nature is only vaguely revealed even microscopically in

Go to **Cram101.com** for the Practice Tests for this Chapter.

Go to **Cram101.com** for the Practice Tests for this Chapter.
And, **NEVER** highlight a book again!

thin section by transmitted polarized light.

Mesozoic	The Mesozoic is one of three geologic eras of the Phanerozoic eon. The Mesozoic was a time of tectonic, climatic and evolutionary activity, shifting from a state of connectedness into their present configuration. The climate was exceptionally warm throughout the period, also playing an important role in the evolution and diversification of new animal species. By the end of the era, the basis of modern life was in place.
Cenozoic	The Cenozoic Era meaning "new life", is the most recent of the three classic geological eras. It covers the 65.5 million years since the Cretaceous-Tertiary extinction event at the end of the Cretaceous that marked the demise of the last non-avian dinosaurs and the end of the Mesozoic Era. The Cenozoic era is ongoing.
Silica	Silica is the oxide of silicon, chemical formula SiO_2, and is known for its hardness as early as the 16th century. It is a principle component in most types of glass and substances such as concrete.
Organism	In biology and ecology, an organism is a living complex adaptive system of organs that influence each other in such a way that they function in some way as a stable whole.
Glauconite	Glauconite is a phyllosilicate mineral. It crystallizes with monoclinic geometry. The name is derived from the Greek glaucos meaning 'gleaming face=symbol>¢ or 'silvery', to describe the appearance of the blue-green color, presumably relating to the sheen and blue-green color of the sea's surface. Its color ranges from olive green, black green to bluish green. It is probably the result of the iron content of the mineral.
Unconformity	An unconformity is a buried erosion surface separating two rock masses or strata of different ages, indicating that sediment deposition was not continuous. In general, the older layer was exposed to erosion for an interval of time before deposition of the younger, but the term is used to describe any break in the sedimentary geologic record.
Mudstone	Mudstone is a fine-grained sedimentary rock whose original constituents were clays or muds. Grain size is up to 0.0625 mm with individual grains too small to be distinguished without a microscope.
Weathering	Weathering is the process of breaking down rocks, soils and their minerals through direct contact with the atmosphere. Weathering occurs without movement. Two main classifications of weathering processes exist. Mechanical or physical weathering involves the breakdown of rocks and soils through direct contact with atmospheric conditions. The second classification, chemical weathering, involves the direct effect of atmospheric chemicals in the breakdown of rocks, soils and minerals.
Placer	A placer is an accumulation of alluvium or eluvium containing valuable minerals which is formed by deposition of dense mineral phases in a trap site.
Sandstone	Sandstone is a sedimentary rock composed mainly of sand-size mineral or rock grains. Most sandstone is composed of quartz and/or feldspar because these are the most common minerals in the Earth's crust. Like sand, sandstone may be any color, but the most common colors are tan, brown, yellow, red, gray and white.
Stream	A stream is a body of water with a current, confined within a bed and banks. Streams are important as conduits in the water cycle, instruments in aquifer recharge, and corridors for fish and wildlife migration.
Magnetite	Magnetite is a ferrimagnetic mineral one of several iron oxides and a member of the spinel group. The chemical IUPAC name is iron oxide and the common chemical name ferrous-ferric oxide.

Go to **Cram101.com** for the Practice Tests for this Chapter.
And, **NEVER** highlight a book again!

Kerogen	Kerogen is a mixture of organic chemical compounds that make up a portion of the organic matter in sedimentary rocks. It is insoluble in normal organic solvents because of the huge molecular weight of its component compounds. The soluble portion is known as bitumen.
Hematite	Hematite is a very common mineral, colored black to steel or silver-gray, brown to reddish brown, or red. It is mined as the main ore of iron. Varieties include kidney ore, martite iron rose and specularite. While the forms of it vary, they all have a rust-red streak. it is harder than pure iron, but much more brittle.
Laterite	Laterite is a surface formation in hot and wet tropical areas which is enriched in iron and aluminium and develops by intensive and long lasting weathering of the underlying parent rock. Nearly all kinds of rocks can be deeply decomposed by the action of high rainfall and elevated temperatures. This gives rise to a residual concentration of more insoluble elements.
Chemical weathering	Chemical weathering involves the change in the composition of rock, often leading to a face=symbol>¢break down' in its form.
Climate	Climate is the average and variations of weather over long periods of time. Climate zones can be defined using parameters such as temperature and rainfall.
Carbon	Carbon is a chemical element. An abundant nonmetallic, tetravalent element, carbon has several allotropic forms. This element is the basis of the chemistry of all known life.
Peat	Peat is an accumulation of partially decayed vegetation matter. It forms in wetlands.
Coal	Coal is a fossil fuel formed in swamp ecosystems where plant remains were saved by water and mud from oxidization and biodegradation. It is a sedimentary rock, but the harder forms, such as anthracite coal, can be regarded as metamorphic rocks because of later exposure to elevated temperature and pressure. It is composed primarily of carbon along with assorted other elements, including sulfur.
Lignite	Lignite is the lowest rank of coal and used almost exclusively as fuel for steam-electric power generation.
Bituminous coal	Bituminous coal is a relatively hard coal containing a tar-like substance called bitumen. It is of higher quality than lignite coal but of poorer quality than anthracite coal.
Conchoidal fracture	Conchoidal fracture describes the way that brittle materials break when they do not follow any natural planes of separation. Materials that break in this way include flint and other fine-grained minerals, as well as most amorphous solids, such as obsidian and other types of glass.
Anthracite	Anthracite is a hard, compact variety of mineral coal that has a high luster. It has the highest carbon count and contains the fewest impurities of all coals, despite its lower calorific content.
Oil shale	Oil shale is a general term applied to a fine-grained sedimentary rock containing significant traces of kerogen that have not been buried for sufficient time to produce conventional fossil fuels. When heated to a sufficiently high temperature a vapor is driven off which can be distilled to yield a petroleum.
Magma	Magma is molten rock located beneath the surface of the Earth, and which often collects in a magma chamber. Magma is a complex high-temperature fluid substance. Most are silicate solutions. It is capable of intrusion into adjacent rocks or of extrusion onto the surface as lava or ejected explosively as tephra to form pyroclastic rock. Environments of magma formation include subduction zones, continental rift zones, mid-oceanic ridges, and hotspots, some of which are interpreted as mantle plumes.
Silicate	In geology and astronomy, the term silicate is used to denote types of rock that consist

Go to **Cram101.com** for the Practice Tests for this Chapter.
And, **NEVER** highlight a book again!

predominantly of silicate minerals. Such rocks include a wide range of igneous, metamorphic and sedimentary types. Most of the Earth's mantle and crust are made up of silicate rocks. The same is true of the Moon and the other rocky planets.

Metamorphic rock

Metamorphic rock is the result of the transformation of a pre-existing rock type, the protolith, in a process called metamorphism, which means "change in form". The protolith is subjected to heat and extreme pressure causing profound physical and/or chemical change. The protolith may be sedimentary rock, igneous rock or another older rock.

Metamorphic rocks

Metamorphic rock is the result of the transformation of a pre-existing rock type, the protolith, in a process called metamorphism. The protolith is subjected to heat and extreme pressure causing profound physical and/or chemical change. Metamorphic rocks make up a large part of the Earth's crust. They are formed deep beneath the Earth's surface by great stresses from rocks above and high pressures and temperatures.

Metamorphism

Metamorphism can be defined as the solid state recrystallisation of pre-existing rocks due to changes in heat and/or pressure and/or introduction of fluids. There will be mineralogical, chemical and crystallographic changes. Metamorphism produced with increasing pressure and temperature conditions is known as prograde metamorphism. Conversely, decreasing temperatures and pressure characterize retrograde metamorphism.

Igneous

Igneous rocks form when molten rock, magma, cools and solidifies, with or without crystallization, either below the surface as intrusive, plutonic rocks or on the surface as extrusive, volcanic, rocks.

Igneous rock

Igneous rock forms when rock cools and solidifies either below the surface as intrusive rocks or on the surface as extrusive rocks. This magma can be derived from partial melts of pre-existing rocks in either the Earth's mantle or crust. Typically, the melting is caused by one or more of the following processes -- an increase in temperature, a decrease in pressure, or a change in composition.

Crystal

A crystal is a solid in which the constituent atoms, molecules, or ions are packed in a regularly ordered, repeating pattern extending in all three spatial dimensions. Most metals encountered in everyday life are polycrystals. Crystals are often symmetrically intergrown to form crystal twins.

Crystallization

Crystallization is the process of formation of solid crystals from a uniform solution. It is also a chemical solid-liquid separation technique, in which mass transfer of a solute from the liquid solution to a pure solid crystalline phase occurs.

Igneous textures

Igneous Textures This term is applied to igneous rocks in order to describe their appearance. This allows geologist to determine how a particular igneous rock formed and under what conditions it formed. There are six main types of textures; phaneritic, aphanitic, porphyritic, glassy, pyroclastic and pegmatitic.

Granodiorite

Granodiorite is an intrusive igneous rock similar to granite, but contains more plagioclase than potassium feldspar. It usually contains abundant biotite mica and hornblende, giving it a darker appearance than true granite.

Plagioclase

Plagioclase is a very important series of tectosilicate minerals within the feldspar family. Rather than referring to a particular mineral with a specific chemical composition, it is a solid solution series.

Rapid

A rapid is a section of a river of relatively steep gradient causing an increase in water flow and turbulence. A rapid is a hydrological feature between a run and a cascade. It is characterized by the river becoming shallower and having some rocks exposed above the flow surface.

Vapor	Vapor is the gas phase component of a another state of matter which does not completely fill its container. It is distinguished from the pure gas phase by the presence of the same substance in another state of matter. Hence when a liquid has completely evaporated, it is said that the system has been completely transformed to the gas phase.
Obsidian	Obsidian is a type of naturally-occurring glass formed as an extrusive igneous rock. It is produced when felsic lava erupted from a volcano cools rapidly through the glass transition temperature and freezes without sufficient time for crystal growth. Obsidian is commonly found within the margins of rhyolitic lava flows known as obsidian flows, where cooling of the lava is rapid.
Olivine	The mineral olivine is a magnesium iron silicate. It is one of the most common minerals on Earth, and has also been identified on the Moon, Mars, and comet Wild 2.
Pyroxenes	The pyroxenes are a group of important rock-forming silicate minerals found in many igneous and metamorphic rocks. They share a common structure comprised of single chains of silica tetrahedra and they crystalise in the monoclinic and orthorhombic system.
Dike	A dike is an intrusion into a cross-cutting fissure, meaning a dike cuts across other pre-existing layers or bodies of rock, this means that a dike is always younger than the rocks that contain it. The thickness is usually much smaller than the other two dimensions. Thickness can vary from sub-centimeter scale to many meters in thickness and the lateral dimensions can extend over many kilometers.
Volcanic gases	Volcanic gases include a variety of substances given off by active volcanos. These include gases trapped in cavities in volcanic rocks, dissolved or dissociated gases in magma and lava, or gases emanating directly from lava or indirectly through ground water heated by volcanic action.
Tephra	Tephra is air-fall material produced by a volcanic eruption regardless of composition or fragment size. It is typically rhyolitic in composition as most explosive volcanoes are the product of the more viscous felsic or high silica magmas.
Pyroclastics	Pyroclastics are clastic rocks composed solely or primarily of volcanic materials.
Tuff	Tuff is a type of rock consisting of consolidated volcanic ash ejected from vents during a volcanic eruption.
Lapillus	Lapillus refers to a size classification term for tephra, which is material that falls out of the air during a volcanic eruption. They are in some senses similar to ooids or pisoids in calcareous sediments.
Biotite	Biotite is a common phyllosilicate mineral within the mica group. Primarily a solid-solution series between the iron-endmember annite, and the magnesium-endmember phlogopite; more aluminous endmembers include siderophyllite.
Blocks	Blocks in meteorology are large scale patterns in the atmospheric pressure field that are nearly stationary, effectively "blocking" or redirecting migratory cyclones. These blocks can remain in place for several days or even weeks, causing the areas affected by them to have the same kind of weather for an extended period of time.
Mafic	In geology, mafic minerals and rocks are silicate minerals, magmas, and volcanic and intrusive igneous rocks that have relatively high concentrations of the heavier elements. The term is a combination of "magnesium" and ferrum.
Phenocryst	A phenocryst is a relatively large and usually conspicuous crystal distinctly larger than the grains of the rock groundmass of a porphyritic igneous rock. They often have euhedral forms either due to early growth within a magma or by post-emplacement recrystallization.
Groundmass	Groundmass rock is the fine-grained mass of material in which larger grains or crystals are

Go to **Cram101.com** for the Practice Tests for this Chapter.
And, **NEVER** highlight a book again!

embedded. The groundmass of an igneous rock consists of fine-grained, often microscopic, crystals in which larger crystals are embedded. This porphyritic texture is indicative of multi-stage cooling of magma.

Porphyry	Porphyry is a variety of igneous rock consisting of large-grained crystals, such as feldspar or quartz, dispersed in a fine-grained feldspathic matrix or groundmass. The larger crystals are called phenocrysts.
Hornblende	Hornblende is a complex inosilicate series of minerals. Hornblende is not a recognized mineral, in its own right but the name is used as a general or field term, to refer to a dark amphibole. It is an isomorphous mixture of three molecules; a calcium-iron-magnesium silicate, an aluminium-iron-magnesium silicate and an iron-magnesium silicate.
Amygdule	An amygdule forms when the vesicular cavities are filled with a secondary mineral such as calcite, quartz, chlorite or one of the zeolites, which are deposited by having minerals "wash" through the pores in the rock. They are filled from the outside, making some of them concentrically layered.
Basalt	Basalt is a common gray to black extrusive volcanic rock. It is usually fine-grained due to rapid cooling of lava on the Earth's surface. It may be porphyritic containing larger crystals in a fine matrix, or vesicular, or frothy scoria.
Chlorite	A chlorite is a compound that contains this group, with chlorine in oxidation state +3. They are also known as salts of chlorous acid.
Orthoclase	Orthoclase is an important tectosilicate mineral, which forms igneous rock. Orthoclase is named based on the Greek for "straight fracture," because its two cleavages are at right angles to each other. Orthoclase crystallizes in the monoclinic crystal system. It has a hardness of 6, a specific gravity of 2.56-2.58, and a vitreous to pearly luster. It can be colored white, gray, yellow, pink, or red; rarely green.
Amphibole	Amphibole defines an important group of generally dark-colored rock-forming inosilicate minerals linked at the vertices and generally containing ions of iron and/or magnesium in their structures. Amphiboles crystallize into two crystal systems, monoclinic and orthorhombic.
Andesite	Andesite is an igneous, volcanic rock, of intermediate composition, with aphanitic to porphyritic texture.
Diorite	Diorite is a grey to dark grey intermediate intrusive igneous rock composed principally of plagioclase feldspar, biotite, hornblende, and/or pyroxene. It may contain small amounts of quartz, microcline and olivine.
Gabbro	Gabbro is a dark, coarse-grained, intrusive igneous rock chemically equivalent to basalt. It is a plutonic rock, formed when molten magma is trapped beneath the Earth face=symbol>¢s surface and cools into a crystalline mass.
Granite	Granite is a common and widely occurring type of intrusive, felsic, igneous rock. Granites are usually medium to coarsely crystalline, occasionally with some individual crystals larger than the groundmass forming a rock known as porphyry. Granites can be pink to dark gray or even black, depending on their chemistry and mineralogy.
Microcline	Microcline is an important igneous rock forming tectosilicate mineral. It is common in granite and pegmatites. Microcline forms during slow cooling of orthoclase; it is stable at lower temperatures than orthoclase.
Fold	The term fold is used in geology when one or a stack of originally flat and planar surfaces, such as sedimentary strata, are bent or curved as a result of plastic, i.e. permanent, deformation.

Foliated metamorphic rock	Foliated metamorphic rock has penetrative planar fabric present within it. It is common to rocks affected by regional metamorphic compression typical of orogenic belts.
Mylonite	Mylonite is a fine-grained, compact rock produced by dynamic crystallization of the constituent minerals resulting in a reduction of the grain size of the rock. It is classified as a metamorphic rock. They can have many different mineralogical compositions, it is a classification based on the textural appearance of the rock.
Facies	A facies should ideally be a distinctive rock that forms under certain conditions of sedimentation, reflecting a particular process or environment.

Alcatraz Island	Alcatraz Island is a small island located in the middle of San Francisco Bay in California, United States. It served as a lighthouse, then a military fortification, then a military prison followed by a federal prison until 1963, when it became a national recreation area.
Ridge	A ridge is a geological feature that is also known as a Rip in the earth causing magma to flow out and forming an undersea volcano, it also has geological features, a continuous elevational crest for some distance. Ridges are usually termed hills or mountains as well, depending on size.
Fold	The term fold is used in geology when one or a stack of originally flat and planar surfaces, such as sedimentary strata, are bent or curved as a result of plastic, i.e. permanent, deformation.
Creep	Creep, is the slow downward progression of rock and soil down a low grade slope; it can also refer to slow deformation of such materials as a result of prolonged pressure and stress.
Concretion	A concretion is a volume of sedimentary rock in which a mineral cement fills the porosity. They are often ovoid or spherical in shape, although irregular shapes also occur. They form within layers of sedimentary strata that have already been deposited. They usually form early in the burial history of the sediment, before the rest of the sediment is hardened into rock.
Fossil	Fossils are the mineralized or otherwise preserved remains or traces of animals, plants, and other organisms. The totality of fossils, both discovered and undiscovered, and their placement in fossiliferous rock formations and sedimentary layers is known as the fossil record.
Granite	Granite is a common and widely occurring type of intrusive, felsic, igneous rock. Granites are usually medium to coarsely crystalline, occasionally with some individual crystals larger than the groundmass forming a rock known as porphyry. Granites can be pink to dark gray or even black, depending on their chemistry and mineralogy.
Mineral	A mineral is a naturally occurring substance formed through geological processes that has a characteristic chemical composition, a highly ordered atomic structure and specific physical properties. A rock, by comparison, is an aggregate of minerals and need not have a specific chemical composition. Minerals range in composition from pure elements and simple salts to very complex silicates with thousands of known forms.
Quartz	Quartz is the second most common mineral in the Earth's continental crust. It is made up of a lattice of silica tetrahedra. Quartz belongs to the rhombohedral crystal system. In nature quartz crystals are often twinned, distorted, or so intergrown with adjacent crystals of quartz or other minerals as to only show part of this shape, or to lack obvious crystal faces altogether and appear massive.
Limestone	Limestone is a sedimentary rock composed largely of the mineral calcite. Limestone often contains variable amounts of silica in the form of chert or flint, as well as varying amounts of clay, silt and sand as disseminations, nodules, or layers within the rock. The primary source of the calcite in limestone is most commonly marine organisms. These organisms secrete shells that settle out of the water column and are deposited on ocean floors as pelagic ooze or alternatively is conglomerated in a coral reef.
Gabbro	Gabbro is a dark, coarse-grained, intrusive igneous rock chemically equivalent to basalt. It is a plutonic rock, formed when molten magma is trapped beneath the Earth face=symbol>¢s surface and cools into a crystalline mass.
Shale	Shale is a fine-grained sedimentary rock whose original constituents were clays or muds. It is characterized by thin laminae breaking with an irregular curving fracture, often splintery and usually parallel to the often-indistinguishable bedding plane.
Sandstone	Sandstone is a sedimentary rock composed mainly of sand-size mineral or rock grains. Most

Go to **Cram101.com** for the Practice Tests for this Chapter.

sandstone is composed of quartz and/or feldspar because these are the most common minerals in the Earth's crust. Like sand, sandstone may be any color, but the most common colors are tan, brown, yellow, red, gray and white.

Unconformity	An unconformity is a buried erosion surface separating two rock masses or strata of different ages, indicating that sediment deposition was not continuous. In general, the older layer was exposed to erosion for an interval of time before deposition of the younger, but the term is used to describe any break in the sedimentary geologic record.
Dolomite	Dolomite is the name of a sedimentary carbonate rock and a mineral, both composed of calcium magnesium carbonate found in crystals. Dolomite rock is composed predominantly of the mineral dolomite. Limestone that is partially replaced by dolomite is referred to as dolomitic limestone.
Season	A season is one of the major divisions of the year, generally based on yearly periodic changes in weather. They are recognized as: spring, summer, autumn, and winter.
Stratigraphy	Stratigraphy, a branch of geology, studies rock layers and layering. It is primarily used in the study of sedimentary and layered volcanic rocks. Stratigraphy includes two related subfields: lithologic or lithostratigraphy and biologic stratigraphy or biostratigraphy.
Electron correlation	Electron correlation refers to the interaction between electrons in a quantum system whose electronic structure is being considered.
Fault	Faults are planar rock fractures, which show evidence of relative movement. Large faults within the Earth's crust are the result of shear motion and active fault zones are the causal locations of most earthquakes. Earthquakes are caused by energy release during rapid slippage along faults. The largest examples are at tectonic plate boundaries but many faults occur far from active plate boundaries. Since faults do not usually consist of a single, clean fracture, the term fault zone is used when referring to the zone of complex deformation that is associated with the fault plane.
Foliated metamorphic rock	Foliated metamorphic rock has penetrative planar fabric present within it. It is common to rocks affected by regional metamorphic compression typical of orogenic belts.
Stream	A stream is a body of water with a current, confined within a bed and banks. Streams are important as conduits in the water cycle, instruments in aquifer recharge, and corridors for fish and wildlife migration.
Excavation	Excavation is the most commonly used technique within the science of archaeology. It is the exposure, processing and recording of archaeological remains.
Striations	In geology, glacial striations are grooves or lines inscribed on the surface of a rock, produced by a geological process such as glacial flow.
Terrain	Terrain is the third or vertical dimension of land surface. When terrain is described underwater, the term bathymetry is used.
Bedrock	Bedrock is the native consolidated rock underlying the Earth's surface. Above the bedrock is usually an area of broken and weathered unconsolidated rock in the basal subsoil.
Well logs	Well logs is a technique used in the oil and gas industry for recording rock and fluid properties to find hydrocarbon zones in the geological formations within the Earth face=symbol>¢s crust.
Pipeline	Pipeline transport is a transportation of goods through a pipe. Most commonly, liquid and gases are sent, but pneumatic tubes that transport solid capsules using compressed air have also been used..

Loess	Among the classifications of soil types, loess, is a fine, silty, windblown type of unconsolidated deposit. It is derived from glacial deposits, where glacial activity has ground rocks very fine. After drying, these deposits are highly susceptible to wind erosion, and downwind deposits may become very deep. Loess deposits are geologically unstable by nature, and will erode even without being disturbed by humans.
Paleosol	Paleosol can refer to two meanings. The first meaning, is simply that of a former soil preserved by burial underneath either sediments or volcanic deposits, which in case of older deposits, have lithified into rock. In general, it is the typical and accepted practice to use the term "paleosol" to designate such "fossil" soils found buried within either sedimentary or volcanic deposits exposed in all continents.
Alluvial soil	Alluvial soil or sediments are deposited by a river or other running water. Alluvial soil is typically made up of a variety of materials, including fine particles of silt and clay and larger particles of sand and gravel.
Silt	Silt is soil or rock derived granular material of a specific grain size. Silt may occur as a soil or alternatively as suspended sediment in a water column of any surface water body. It may also exist as deposition soil at the bottom of a water body.
Triassic	Triassic is the first period of the Mesozoic Era. Both the start and end of the Triassic are marked by major extinction events. During the Triassic, both marine and continental life show an adaptive radiation beginning from the starkly impoverished biosphere that followed the Permian-Triassic extinction. Corals of the hexacorallia group made their first appearance. The first flowering plants may have evolved during the Triassic, as did the first flying vertebrates, the pterosaurs.
Dune	A dune is a hill of sand built by eolian processes. Dunes are subject to different forms and sizes based on their interaction with the wind. Most kinds of dune are longer on the windward side where the sand is pushed up the dune, and a shorter in the lee of the wind. The trough between dunes is called a slack. A "dune field" is an area covered by extensive sand dunes. Large dune fields are known as ergs.
Alluvial	An alluvial plain is a relatively flat and gently sloping landform found at the base of a range of hills or mountains, formed by the deposition of alluvial soil over a long period of time by one or more rivers coming from the mountains.
Porosity	Porosity is a measure of the void spaces in a material, and is measured as a fraction, between 0–1, or as a percentage between 0–100%.
Ductile materials	Ductile materials have the mechanical property of being capable of sustaining large plastic deformations due to tensile stress without fracture in metals, such as being drawn into a wire. It is characterized by the material flowing under shear stress. It is contrasted with brittleness.
Engineering geology	Engineering Geology is the application of the geologic sciences to engineering practice for the purpose of assuring that the geologic factors affecting the location, design, construction, operation and maintenance of engineering works are recognized and adequately provided for.
Geologic map	A geologic map is a special-purpose map made to show geological features. The stratigraphic contour lines are drawn on the surface of a selected deep stratum, so that they can show the topographic trends of the strata under the ground. It is not always possible to properly show this when the strata are extremely fractured, mixed, in some discontinuities, or where they are otherwise disturbed.
Geology	Geology is the science and study of the solid matter that constitute the Earth. Encompassing such things as rocks, soil, and gemstones, geology studies the composition, structure,

Go to **Cram101.com** for the Practice Tests for this Chapter.

physical properties, history, and the processes that shape Earth's components.

Pollution	Pollution is the introduction of substances or energy into the environment, resulting in deleterious effects of such a nature as to endanger human health, harm living resources and ecosystems, and impair or interfere with amenities and other legitimate uses of the environment.
Fauna	Fauna is a collective term for animal life of any particular region or time. Paleontologists usually use fauna to refer to a typical collection of animals found in a specific time or place. Paleontologists sometimes refer to a sequence of 80 or so faunal stages, which are a series of rocks all containing similar fossils.
Air pollution	Air Pollution is a chemical, physical, or biological agent that modifies the natural characteristics of the atmosphere. The atmosphere is a complex, dynamic natural gaseous system that is essential to support life on planet Earth. Stratospheric ozone depletion due to air pollution has long been recognized as a threat to human health as well as to the Earth's ecosystems. Worldwide air pollution is responsible for large numbers of deaths and cases of respiratory disease.
Rapid	A rapid is a section of a river of relatively steep gradient causing an increase in water flow and turbulence. A rapid is a hydrological feature between a run and a cascade. It is characterized by the river becoming shallower and having some rocks exposed above the flow surface.
Landslide	A landslide is a geological phenomenon which includes a wide range of ground movement, such as rock falls, deep failure of slopes and shallow debris flows. Although gravity face=symbol>¢s action on an over-steepened slope is the primary reason for a landslide, there are other contributing factors affecting the original slope stability.

Go to **Cram101.com** for the Practice Tests for this Chapter.

Topographic maps	Topographic maps are a variety of maps characterized by large-scale detail and quantitative representation of relief, usually using contour lines in modern mapping, but historically using a variety of methods.
Sea level	Mean sea level is the average height of the sea, with reference to a suitable reference surface.
Landform	A landform comprises a geomorphological unit, and is largely defined by its surface form and location in the landscape, as part of the terrain, and as such, is typically an element of topography. They are categorised by features such as elevation, slope, orientation, stratification, rock exposure, and soil type. They include berms, mounds, hills, cliffs, valleys, rivers and numerous other elements.
Contour interval	In cartography, a contour interval is any space between vertical lines on a topographic map or globe, representing a difference in elevation between the lines.
Longitude	Longitude is the east-west geographic coordinate measurement most commonly utilized in cartography and global navigation.
Quadrangle	In geology or geography, the word quadrangle usually refers to a United States Geological Survey 7.5-minute map, which are usually named after a local physiographic feature.
Latitude	Latitude gives the location of a place on Earth north or south of the equator. Lines of Latitude are the horizontal lines shown running east-to-west on maps. Technically, Latitude is an angular measurement in degrees ranging from 0° at the Equator to 90° at the poles.
Terrain	Terrain is the third or vertical dimension of land surface. When terrain is described underwater, the term bathymetry is used.
Contour line	A contour line shows elevation. A contour line for a function of two variables is a curve connecting points where the function has a same particular value. A contour map is a map illustrated with contour lines, for example a topographic map.
Season	A season is one of the major divisions of the year, generally based on yearly periodic changes in weather. They are recognized as: spring, summer, autumn, and winter.
Stream	A stream is a body of water with a current, confined within a bed and banks. Streams are important as conduits in the water cycle, instruments in aquifer recharge, and corridors for fish and wildlife migration.
Stream gradient	Stream gradient is the ratio of drop in a stream per unit distance, usually expressed as feet per mile or meters per kilometer. A high gradient indicates a steep slope and rapid flow of water; whereas a low gradient indicates a more nearly level stream bed and sluggishly moving water, that may be able to carry only small amounts of very fine sediment.
Fold	The term fold is used in geology when one or a stack of originally flat and planar surfaces, such as sedimentary strata, are bent or curved as a result of plastic, i.e. permanent, deformation.
Fault	Faults are planar rock fractures, which show evidence of relative movement. Large faults within the Earth's crust are the result of shear motion and active fault zones are the causal locations of most earthquakes. Earthquakes are caused by energy release during rapid slippage along faults. The largest examples are at tectonic plate boundaries but many faults occur far from active plate boundaries. Since faults do not usually consist of a single, clean fracture, the term fault zone is used when referring to the zone of complex deformation that is associated with the fault plane.
Altitude	Altitude is the elevation of an object from a known level or datum. Common datums are mean sea level and the surface of the World Geodetic System geoid, used by Global Positioning

	System. In aviation, altitude is measured in feet. For non-aviation uses, altitude may be measured in other units such as metres or miles.
Canyon	A canyon is a deep valley between cliffs often carved from the landscape by a river. Most were formed by a process of long-time erosion from a plateau level. The cliffs form because harder rock strata that are resistant to erosion and weathering remain exposed on the valley walls.
Vegetation	Vegetation is a general term for the plant life of a region; it refers to the ground cover provided by plants, and is, by far, the most abundant biotic element of the biosphere. Primeval redwood forests, coastal mangrove stands, sphagnum bogs, desert soil crusts, roadside weed patches, wheat fields, cultivated gardens and lawns; are all encompassed by the term vegetation.
Infrared	Infrared is electromagnetic radiation of a wavelength longer than that of visible light, but shorter than that of radio waves. The name means "below red", red being the color of visible light with the longest wavelength. Infrared has wavelengths between about 750 nm and 1 mm, spanning three orders of magnitude.
Wavelength	In physics, wavelength is the distance between repeating units of a propagating wave of a given frequency. It is commonly designated by the Greek letter lambda. Examples of wave-like phenonomena are light, water waves, and sound waves. Wavelength of a sine wave.In a wave, a property varies with the position.
Thermal	A thermal column is a column of rizing air in the lower altitudes of the Earth face=symbol>¢s atmosphere. Thermals are created by the uneven heating of the Earth's surface from solar radiation, and are an example of convection. The Sun warms the ground, which in turn warms the air directly above it.
Radiation	Radiation as used in physics, is energy in the form of waves or moving subatomic particles.
Forest	A forest is an area with a high density of trees, historically, a wooded area set aside for hunting. These plant communities cover large areas of the globe and function as animal habitats, hydrologic flow modulators, and soil conservers, constituting one of the most important aspects of the Earth's biosphere.

Go to **Cram101.com** for the Practice Tests for this Chapter.

Season	A season is one of the major divisions of the year, generally based on yearly periodic changes in weather. They are recognized as: spring, summer, autumn, and winter.
Topography	Topography is the study of Earth's surface features or those of other planets, moons, and asteroids
Ridge	A ridge is a geological feature that is also known as a Rip in the earth causing magma to flow out and forming an undersea volcano, it also has geological features, a continuous elevational crest for some distance. Ridges are usually termed hills or mountains as well, depending on size.
Temperate	In geography, temperate latitudes of the globe lie between the tropics and the polar circles. The changes in these regions between summer and winter are generally subtle: warm or cool, rather than extreme hot or cold.
Forest	A forest is an area with a high density of trees, historically, a wooded area set aside for hunting. These plant communities cover large areas of the globe and function as animal habitats, hydrologic flow modulators, and soil conservers, constituting one of the most important aspects of the Earth's biosphere.
Foliated metamorphic rock	Foliated metamorphic rock has penetrative planar fabric present within it. It is common to rocks affected by regional metamorphic compression typical of orogenic belts.
Permeability	In the earth sciences, permeability is a measure of the ability of a material to transmit fluids. It is of great importance in determining the flow characteristics of hydrocarbons in oil and gas reservoirs, and of groundwater in aquifers.
Metamorphic rock	Metamorphic rock is the result of the transformation of a pre-existing rock type, the protolith, in a process called metamorphism, which means "change in form". The protolith is subjected to heat and extreme pressure causing profound physical and/or chemical change. The protolith may be sedimentary rock, igneous rock or another older rock.
Metamorphic rocks	Metamorphic rock is the result of the transformation of a pre-existing rock type, the protolith, in a process called metamorphism. The protolith is subjected to heat and extreme pressure causing profound physical and/or chemical change. Metamorphic rocks make up a large part of the Earth's crust. They are formed deep beneath the Earth's surface by great stresses from rocks above and high pressures and temperatures.
Metamorphism	Metamorphism can be defined as the solid state recrystallisation of pre-existing rocks due to changes in heat and/or pressure and/or introduction of fluids. There will be mineralogical, chemical and crystallographic changes. Metamorphism produced with increasing pressure and temperature conditions is known as prograde metamorphism. Conversely, decreasing temperatures and pressure characterize retrograde metamorphism.
Sedimentary	Sedimentary rock is one of the three main rock groups. Rock formed from these covers 75% of the Earth's land area, and includes common types such as chalk, limestone, dolomite, sandstone, and shale.
Vegetation	Vegetation is a general term for the plant life of a region; it refers to the ground cover provided by plants, and is, by far, the most abundant biotic element of the biosphere. Primeval redwood forests, coastal mangrove stands, sphagnum bogs, desert soil crusts, roadside weed patches, wheat fields, cultivated gardens and lawns; are all encompassed by the term vegetation.
Sedimentary rock	Sedimentary rock is one of the three main rock groups. Sedimentary rock covers 75% of the Earth's land area. Four basic processes are involved in the formation of a clastic sedimentary rock: weathering caused mainly by friction of waves,

Go to **Cram101.com** for the Practice Tests for this Chapter.

transportation where the sediment is carried along by a current, deposition and compaction where the sediment is squashed together to form a rock of this kind.

Drainage	Drainage is the natural or artificial removal of surface and sub-surface water from a given area. Many agricultural soils need drainage to improve production or to manage water supplies.
Clay	Clay is a term used to describe a group of hydrous aluminium phyllosilicate minerals, that are typically less than 2 micrometres in diameter. Clay consists of a variety of phyllosilicate minerals rich in silicon and aluminium oxides and hydroxides which include variable amounts of structural water. Clays are generally formed by the chemical weathering of silicate-bearing rocks by carbonic acid but some are formed by hydrothermal activity.
Tuff	Tuff is a type of rock consisting of consolidated volcanic ash ejected from vents during a volcanic eruption.
Sandstone	Sandstone is a sedimentary rock composed mainly of sand-size mineral or rock grains. Most sandstone is composed of quartz and/or feldspar because these are the most common minerals in the Earth's crust. Like sand, sandstone may be any color, but the most common colors are tan, brown, yellow, red, gray and white.
Precipitation	Precipitation is any product of the condensation of atmospheric water vapor that is deposited on the earth's surface. It occurs when the atmosphere becomes saturated with water vapour and the water condenses and falls out of solution. Air becomes saturated via two processes, cooling and adding moisture.
Conglomerate	A conglomerate is a rock consisting of individual stones that have become cemented together. Conglomerates are sedimentary rocks consisting of rounded fragements and are thus differentiated from breccias, which consist of angular clasts. Both conglomerates and breccias are characterized by clasts larger than sand.
Sinkhole	A sinkhole is a natural depression or hole in the surface topography caused by the removal of soil or bedrock, often both, by water. They may vary in size from less than a meter to several hundred meters both in diameter and depth, and vary in form from soil-lined bowls to bedrock-edged chasms.
Limestone	Limestone is a sedimentary rock composed largely of the mineral calcite. Limestone often contains variable amounts of silica in the form of chert or flint, as well as varying amounts of clay, silt and sand as disseminations, nodules, or layers within the rock. The primary source of the calcite in limestone is most commonly marine organisms. These organisms secrete shells that settle out of the water column and are deposited on ocean floors as pelagic ooze or alternatively is conglomerated in a coral reef.
Gypsum	Gypsum is a very soft mineral composed of calcium sulfate dihydrate, with the chemical formula $CaSO_4 \cdot 2H_2O$. Gypsum occurs in nature as flattened and often twinned crystals and transparent cleavable masses. It may also occur silky and fibrous. Finally it may also be granular or quite compact.
Halite	Halite is the mineral form of sodium chloride. Halite forms isometric crystals. It commonly occurs with other evaporite deposit minerals such as several of the sulfates, halides and borates. Halite occurs in vast lakes of sedimentary evaporite minerals that result from the drying up of enclosed beds, playas, and seas.
Tertiary	The Tertiary covers roughly the time span between the demise of the non-avian dinosaurs and beginning of the most recent Ice Age. Each epoch of the Tertiary was marked by striking developments in mammalian life. The earliest recognizable hominoid relatives of humans appeared. Tectonic activity continued as Gondwana finally split completely apart.
Valley	In geology, a valley is a depression with predominant extent in one direction. The terms U-

	shaped and V-shaped are descriptive terms of geography to characterize the form of valleys. Most valleys belong to one of these two main types or a mixture of them, at least with respect of the cross section of the slopes or hillsides.
Dike	A dike is an intrusion into a cross-cutting fissure, meaning a dike cuts across other pre-existing layers or bodies of rock, this means that a dike is always younger than the rocks that contain it. The thickness is usually much smaller than the other two dimensions. Thickness can vary from sub-centimeter scale to many meters in thickness and the lateral dimensions can extend over many kilometers.
Scree	Scree is a term given to broken rock that appears at the bottom of crags, mountain cliffs or valley shoulders, forming a scree slope. The maximum inclination of such deposits corresponds to the angle of repose of the mean debris size.
Landform	A landform comprises a geomorphological unit, and is largely defined by its surface form and location in the landscape, as part of the terrain, and as such, is typically an element of topography. They are categorised by features such as elevation, slope, orientation, stratification, rock exposure, and soil type. They include berms, mounds, hills, cliffs, valleys, rivers and numerous other elements.
Lava	Lava is molten rock expelled by a volcano during an eruption. When first extruded from a volcanic vent, it is a liquid at temperatures from 700 °C to 1,200 °C.
Terrain	Terrain is the third or vertical dimension of land surface. When terrain is described underwater, the term bathymetry is used.
Topographic maps	Topographic maps are a variety of maps characterized by large-scale detail and quantitative representation of relief, usually using contour lines in modern mapping, but historically using a variety of methods.
Erosion	Erosion is displacement of solids by the agents of ocean currents, wind, water, or ice by downward or down-slope movement in response to gravity or by living organisms.
Weathering	Weathering is the process of breaking down rocks, soils and their minerals through direct contact with the atmosphere. Weathering occurs without movement. Two main classifications of weathering processes exist. Mechanical or physical weathering involves the breakdown of rocks and soils through direct contact with atmospheric conditions. The second classification, chemical weathering, involves the direct effect of atmospheric chemicals in the breakdown of rocks, soils and minerals.
Fault	Faults are planar rock fractures, which show evidence of relative movement. Large faults within the Earth's crust are the result of shear motion and active fault zones are the causal locations of most earthquakes. Earthquakes are caused by energy release during rapid slippage along faults. The largest examples are at tectonic plate boundaries but many faults occur far from active plate boundaries. Since faults do not usually consist of a single, clean fracture, the term fault zone is used when referring to the zone of complex deformation that is associated with the fault plane.
Active fault	An active fault is a fault which has had displacement or seismic activity during the geologically recent period. In the United States, an active fault is generally defined as a fault which displaced earth materials during the Holocene Epoch . An active fault is the most common source of earthquakes and tectonic movements.
Petrology	Petrology is a field of geology which focuses on the study of rocks and the conditions by which they form. There are three branches of petrology, corresponding to the three types of rocks: igneous, metamorphic, and sedimentary. Petrology utilizes the classical fields of mineralogy, petrography, optical mineralogy, and chemical analyses to describe the composition and texture of rocks.

Bedrock	Bedrock is the native consolidated rock underlying the Earth's surface. Above the bedrock is usually an area of broken and weathered unconsolidated rock in the basal subsoil.
Geology	Geology is the science and study of the solid matter that constitute the Earth. Encompassing such things as rocks, soil, and gemstones, geology studies the composition, structure, physical properties, history, and the processes that shape Earth's components.

Topography	Topography is the study of Earth's surface features or those of other planets, moons, and asteroids
Telescope	A telescope is an instrument designed for the observation of remote objects. The term usually refers to optical telescopes, but there are telescopes for most of the spectrum of electromagnetic radiation and for other signal types.
Thermal	A thermal column is a column of rizing air in the lower altitudes of the Earth face=symbol>¢s atmosphere. Thermals are created by the uneven heating of the Earth's surface from solar radiation, and are an example of convection. The Sun warms the ground, which in turn warms the air directly above it.
Island	An island is any piece of land that is completely surrounded by water, above high tide. There are two main types of islands: continental islands and oceanic islands. There are also artificial islands. A grouping of geographically and/or geologically related islands is called an archipelago.
Weather	The weather is the set of all extant phenomena in a given atmosphere at a given time. The term usually refers to the activity of these phenomena over short periods, as opposed to the term climate, which refers to the average atmospheric conditions over longer periods of time.
Theodolite	A theodolite is an instrument for measuring both horizontal and vertical angles, as used in triangulation networks. It is a key tool in surveying and engineering work, but they have been adapted for other specialized purposes in fields like meteorology and rocket launch technology.
Season	A season is one of the major divisions of the year, generally based on yearly periodic changes in weather. They are recognized as: spring, summer, autumn, and winter.
Terrain	Terrain is the third or vertical dimension of land surface. When terrain is described underwater, the term bathymetry is used.
Ridge	A ridge is a geological feature that is also known as a Rip in the earth causing magma to flow out and forming an undersea volcano, it also has geological features, a continuous elevational crest for some distance. Ridges are usually termed hills or mountains as well, depending on size.
Stream	A stream is a body of water with a current, confined within a bed and banks. Streams are important as conduits in the water cycle, instruments in aquifer recharge, and corridors for fish and wildlife migration.
Geology	Geology is the science and study of the solid matter that constitute the Earth. Encompassing such things as rocks, soil, and gemstones, geology studies the composition, structure, physical properties, history, and the processes that shape Earth's components.

Sedimentary	Sedimentary rock is one of the three main rock groups. Rock formed from these covers 75% of the Earth's land area, and includes common types such as chalk, limestone, dolomite, sandstone, and shale.
Sediment	Sediment is any particulate matter that can be transported by fluid flow and which eventually is deposited as a layer of solid particles on the bed or bottom of a body of water or other liquid.
Currents	Ocean currents are any more or less continuous, directed movement of ocean water that flows in one of the Earth's oceans. They are rivers of hot or cold water within the ocean. They are generated from the forces acting upon the water like the earth's rotation, the wind, the temperature and salinity differences and the gravitation of the moon.
Geometry	Geometry is a part of mathematics concerned with questions of size, shape, and relative position of figures and with properties of space. Geometry is one of the oldest sciences. Initially a body of practical knowledge concerning lengths, areas, and volumes, in the third century B.C. geometry was put into an axiomatic form by Euclid, whose treatment set a standard for many centuries to follow.
Particle size	Particle size refers to the diameter of individual grains of sediment, or the lithified particles in clastic rocks. The term may also be applied to other granular materials. This is different from the crystallite size, which is the size of a single crystal inside the particles or grains.
Wave	A wave is a disturbance that propagates through space or spacetime, transferring energy and momentum and sometimes angular momentum.
Stream	A stream is a body of water with a current, confined within a bed and banks. Streams are important as conduits in the water cycle, instruments in aquifer recharge, and corridors for fish and wildlife migration.
Velocity	In physics, velocity is defined as the rate of change of displacement or the rate of displacement. Simply put, it is distance per units of time.
Dune	A dune is a hill of sand built by eolian processes. Dunes are subject to different forms and sizes based on their interaction with the wind. Most kinds of dune are longer on the windward side where the sand is pushed up the dune, and a shorter in the lee of the wind. The trough between dunes is called a slack. A "dune field" is an area covered by extensive sand dunes. Large dune fields are known as ergs.
Ridge	A ridge is a geological feature that is also known as a Rip in the earth causing magma to flow out and forming an undersea volcano, it also has geological features, a continuous elevational crest for some distance. Ridges are usually termed hills or mountains as well, depending on size.
Turbidity	Turbidity is a cloudiness or haziness of water caused by individual particles that are generally invisible to the naked eye, thus being much like smoke in air. Turbidity is generally caused by phytoplankton. Measurement of turbidity is a key test of water quality.
Turbidity current	A turbidity current is a current of rapidly moving, sediment-laden water moving down a slope through air, water, or another fluid. The current moves because it has a higher density and turbidity than the fluid through which it flows.
Rapid	A rapid is a section of a river of relatively steep gradient causing an increase in water flow and turbulence. A rapid is a hydrological feature between a run and a cascade. It is characterized by the river becoming shallower and having some rocks exposed above the flow surface.

Conglomerate	A conglomerate is a rock consisting of individual stones that have become cemented together. Conglomerates are sedimentary rocks consisting of rounded fragements and are thus differentiated from breccias, which consist of angular clasts. Both conglomerates and breccias are characterized by clasts larger than sand.
Deposition	Deposition is the geological process whereby material is added to a landform. This is the process by which wind and water create a sediment deposit, through the laying down of granular material that has been eroded and transported from another geographical location.
Sandstone	Sandstone is a sedimentary rock composed mainly of sand-size mineral or rock grains. Most sandstone is composed of quartz and/or feldspar because these are the most common minerals in the Earth's crust. Like sand, sandstone may be any color, but the most common colors are tan, brown, yellow, red, gray and white.
Angle of repose	The angle of repose is an engineering property of granular materials. The angle of repose is the maximum angle of a stable slope determined by friction, cohesion and the shapes of the particles.
Eolian	Eolian processes pertain to the activity of the winds and more specifically, to the winds' ability to shape the surface of the Earth and other planets.
Clasts	Clastic sedimentary rocks are rocks composed predominantly of broken pieces or clasts of older weathered and eroded rocks.
Clay	Clay is a term used to describe a group of hydrous aluminium phyllosilicate minerals, that are typically less than 2 micrometres in diameter. Clay consists of a variety of phyllosilicate minerals rich in silicon and aluminium oxides and hydroxides which include variable amounts of structural water. Clays are generally formed by the chemical weathering of silicate-bearing rocks by carbonic acid but some are formed by hydrothermal activity.
Mudstone	Mudstone is a fine-grained sedimentary rock whose original constituents were clays or muds. Grain size is up to 0.0625 mm with individual grains too small to be distinguished without a microscope.
Stratigraphy	Stratigraphy, a branch of geology, studies rock layers and layering. It is primarily used in the study of sedimentary and layered volcanic rocks. Stratigraphy includes two related subfields: lithologic or lithostratigraphy and biologic stratigraphy or biostratigraphy.
Erosion	Erosion is displacement of solids by the agents of ocean currents, wind, water, or ice by downward or down-slope movement in response to gravity or by living organisms.
Turbidite	Turbidite geological formations have their origins in turbidity current deposits, deposits from a form of underwater avalanche that are responsible for distributing vast amounts of clastic sediment into the deep ocean.
Grazing	Grazing is feeding on growing herbage, attached algae, or phytoplankton.
Tectonics	Tectonics is a field of study within geology concerned generally with the structures within the crust of the Earth, or other planets, and particularly with the forces and movements that have operated in a region to create these structures.
Gravel	Gravel is rock that is of a certain particle size range. In geology, gravel is any loose rock that is at least two millimeters in its largest dimension and no more than 75 millimeters.
Earthquake	An earthquake is the result from the sudden release of stored energy in the Earth face=symbol>¢s crust that creates seismic waves. At the Earth face=symbol>¢s surface, earthquakes may manifest themselves by a shaking or displacement of the ground. An earthquake is caused by tectonic plates getting stuck and putting a strain on the ground. The strain becomes so great that rocks give way by breaking and sliding along fault planes.

Go to **Cram101.com** for the Practice Tests for this Chapter.

Compaction	Compaction is the process of a material being more closely packed together.
Concretion	A concretion is a volume of sedimentary rock in which a mineral cement fills the porosity. They are often ovoid or spherical in shape, although irregular shapes also occur. They form within layers of sedimentary strata that have already been deposited. They usually form early in the burial history of the sediment, before the rest of the sediment is hardened into rock.
Bedding planes	Bedding planes are where one sedimetary deposit ends and another one begins. The rock is prone to breakage at these points because of the weakness between the layers.
Fossil	Fossils are the mineralized or otherwise preserved remains or traces of animals, plants, and other organisms. The totality of fossils, both discovered and undiscovered, and their placement in fossiliferous rock formations and sedimentary layers is known as the fossil record.
Dike	A dike is an intrusion into a cross-cutting fissure, meaning a dike cuts across other pre-existing layers or bodies of rock, this means that a dike is always younger than the rocks that contain it. The thickness is usually much smaller than the other two dimensions. Thickness can vary from sub-centimeter scale to many meters in thickness and the lateral dimensions can extend over many kilometers.
Shale	Shale is a fine-grained sedimentary rock whose original constituents were clays or muds. It is characterized by thin laminae breaking with an irregular curving fracture, often splintery and usually parallel to the often-indistinguishable bedding plane.
Sill	In geology, a sill is a tabular pluton that has intruded between older layers of sedimentary rock, beds of volcanic lava or tuff, or even along the direction of foliation in metamorphic rock. The term sill is synonymous with concordant intrusive sheet. This means that the sill does not cut across preexisting rocks. Contrast this with dikes.
Laccolith	A laccolith is an igneous intrusion that has been injected between two layers of sedimentary rock.
Fold	The term fold is used in geology when one or a stack of originally flat and planar surfaces, such as sedimentary strata, are bent or curved as a result of plastic, i.e. permanent, deformation.
Fault	Faults are planar rock fractures, which show evidence of relative movement. Large faults within the Earth's crust are the result of shear motion and active fault zones are the causal locations of most earthquakes. Earthquakes are caused by energy release during rapid slippage along faults. The largest examples are at tectonic plate boundaries but many faults occur far from active plate boundaries. Since faults do not usually consist of a single, clean fracture, the term fault zone is used when referring to the zone of complex deformation that is associated with the fault plane.
Meridian	A meridian is an imaginary line on the Earth's surface from the North Pole to the South Pole that connects all locations with a given longitude. The position of a point on the meridian is given by the latitude. Each meridian is perpendicular to all circles of latitude at the intersection points. Each is also the same size, being half of a great circle on the Earth's surface and therefore measuring 20,003.93 km.
Trace fossil	A trace fossil is a structure preserved in sedimentary rocks that record biological activity. While we are most familiar with relatively spectacular, fossilized hard-part remains such as shells and bones, a trace fossil is often less dramatic, but nonetheless very important.
Sedimentary facies	Sedimentary facies are usually further subdivided, for example, you might refer to a "tan, cross-bedded oolitic limestone facies" or a "shale facies". The characteristics of the rock unit come from the depositional environment and original composition. Sedimentary facies

Go to **Cram101.com** for the Practice Tests for this Chapter.

Go to **Cram101.com** for the Practice Tests for this Chapter.
And, **NEVER** highlight a book again!

	reflect depositional environment, each facies being a distinct kind of sediment for that area or environment.
Facies	A facies should ideally be a distinctive rock that forms under certain conditions of sedimentation, reflecting a particular process or environment.
Siltstone	Siltstone is a sedimentary rock which has a composition intermediate in grain size between the coarser sandstones and the finer mudstones and shales.
Limestone	Limestone is a sedimentary rock composed largely of the mineral calcite. Limestone often contains variable amounts of silica in the form of chert or flint, as well as varying amounts of clay, silt and sand as disseminations, nodules, or layers within the rock. The primary source of the calcite in limestone is most commonly marine organisms. These organisms secrete shells that settle out of the water column and are deposited on ocean floors as pelagic ooze or alternatively is conglomerated in a coral reef.
Storm	A storm is any disturbed state of an astronomical body's atmosphere, especially affecting its surface, and strongly implying severe weather. It may be marked by strong wind, thunder and lightning, heavy precipitation, such as ice, or wind transporting some substance through the atmosphere.
Unconformity	An unconformity is a buried erosion surface separating two rock masses or strata of different ages, indicating that sediment deposition was not continuous. In general, the older layer was exposed to erosion for an interval of time before deposition of the younger, but the term is used to describe any break in the sedimentary geologic record.
Glauconite	Glauconite is a phyllosilicate mineral. It crystallizes with monoclinic geometry. The name is derived from the Greek glaucos meaning 'gleaming face=symbol>¢ or 'silvery', to describe the appearance of the blue-green color, presumably relating to the sheen and blue-green color of the sea's surface. Its color ranges from olive green, black green to bluish green. It is probably the result of the iron content of the mineral.
Oxide	An oxide is a chemical compound containing an oxygen atom and other elements. Most of the earth's crust consists of them. They result when elements are oxidized by air.
Nodule	A nodule in petrology or mineralogy is an irregular rounded to spherical concretion. They are typically solid replacement bodies of chert or iron oxides formed during diagenesis of a sedimentary rock.
Manganese	Manganese is a chemical element. Its ions are variously colored, and are used industrially as pigments and as oxidation chemicals. Its ions function as cofactors for a number of enzymes and the element is thus a required trace mineral for all known living organisms. It is a grey-white metal, resembling iron.
Coprolite	Coprolite is the name given to the mineral that results when human or animal dung is fossilized. Coprolite may range in size from the size of a BB all the way up to that of a large appliance.
Carbonate	In organic chemistry, a carbonate is a salt of carbonic acid.
Cementation	Cementation is the process of deposition of dissolved mineral components in the interstices of sediments. It is an important factor in the consolidation of coarse-grained clastic sedimentary rocks such as sandstones, conglomerates, or breccias during diagenesis or lithification. Cementing materials may include silica, carbonates, iron oxides, or clay minerals.
Silt	Silt is soil or rock derived granular material of a specific grain size. Silt may occur as a

Go to **Cram101.com** for the Practice Tests for this Chapter.

	soil or alternatively as suspended sediment in a water column of any surface water body. It may also exist as deposition soil at the bottom of a water body.
Tide	Tide refers to the cyclic rizing and falling of Earth's ocean surface caused by the tidal forces of the Moon and the sun acting on the oceans. They cause changes in the depth of the marine and estuarine water bodies and produce oscillating currents known as tidal streams, making prediction of tides important for coastal navigation.
Marsh	In geography, a marsh is a type of wetland which is subject to almost continuous inundation. Typically it features grasses, rushes, reeds, typhas, sedges, and other herbaceous plants in a context of shallow water. It is different from a swamp, which has a greater proportion of open water surface, and is generally deeper than a it.
Claystone	Claystone is a geological term used to describe a sedimentary rock that is composed primarily of clay-sized particles.
Intertidal	The intertidal zone in marine aquatic environments is the area of the foreshore and seabed that is exposed to the air at low tide and submerged at high tide.
Stromatolite	A stromatolite is defined as "attached, lithified sedimentary growth structure, accretionary away from a point or limited surface of initiation." A variety of stromatolite morphologies exist including conical, stratiform, branching, domal, and columnar types. They face=symbol>¢re commonly thought to have been formed by the trapping, binding, and cementation of sedimentary grains by microorganisms, especially cyanobacteria.
Anhydrite	Anhydrite is a mineral - anhydrous calcium sulfate, $CaSO_4$. It is in the orthorhombic crystal system, with three directions of perfect cleavage parallel to the three planes of symmetry. It is not isomorphous with the orthorhombic barium and strontium sulfates, as might be expected from the chemical formulas.
Detrital	Detrital is a geological term used to describe particles of rock derived from pre-existing rock through processes of weathering and erosion.
Gravitation	Gravitation, in everyday life, is most familiar as the agency that endows objects with weight. Gravitation is responsible for keeping the Earth and the other planets in their orbits around the Sun; for the formation of tides; and for various other phenomena that we observe. Gravitation is also the reason for the very existence of the Earth, the Sun, and most macroscopic objects in the universe; without it, matter would not have coalesced into these large masses, and life, as we know it, would not exist.
Reef	A reef is a rock, sandbar, or other feature lying beneath the surface of the water yet shallow enough to be a hazard to ships. They result from abiotic processes—deposition of sand, wave erosion planning down rock outcrops, and other natural processes.
Organism	In biology and ecology, an organism is a living complex adaptive system of organs that influence each other in such a way that they function in some way as a stable whole.
Breccia	Breccia is a rock composed of angular fragments of rocks or minerals in a matrix, that is a cementing material, that may be similar or different in composition to the fragments.
Intertidal zone	The intertidal zone in marine aquatic environments is the area of the foreshore and seabed that is exposed to the air at low tide and submerged at high tide. Organisms in the intertidal zone are adapted to an environment of harsh extremes. Water is available regularly with the tides but varies from fresh with rain to highly saline and dry salt with drying between tidal inundations.
Stalactite	A stalactite is a type of speleothem that hangs from the ceiling or wall of limestone caves. Stalactites are formed by the deposition of calcium carbonate and other minerals, which is precipitated from mineralized water solutions. The corresponding formation on the floor

Go to **Cram101.com** for the Practice Tests for this Chapter.
And, **NEVER** highlight a book again!

	underneath a stalactite is known as a stalagmite.
Cave	A cave is a natural underground void large enough for a human to enter. Some people suggest that the term 'cave' should only apply to cavities that have some part which does not receive daylight; however, in popular usage, the term includes smaller spaces like a sea cave, rock shelters, and grottos.
Mineral	A mineral is a naturally occurring substance formed through geological processes that has a characteristic chemical composition, a highly ordered atomic structure and specific physical properties. A rock, by comparison, is an aggregate of minerals and need not have a specific chemical composition. Minerals range in composition from pure elements and simple salts to very complex silicates with thousands of known forms.
Foreshore	The foreshore is the part of a beach that is exposed by the low tides and submerged by high tides. This area can include many different types of habitats, including steep rocky cliffs, sandy beaches or vast mudflats. The area can be a narrow strip or can include many meters of shoreline where shallow beach slope interacts with high tidal excursion.
Bioturbation	Bioturbation is the displacement and mixing of sediment particles by benthic fauna or flora. Faunal activities, such as burrowing, ingestion and defecation of sediment grains, construction and maintenance of galleries, and infilling of abandoned dwellings, displace sediment grains and mix the sediment matrix. Bioturbation is a diagenetic process and acts to alter the physical structure, as well as the chemical nature of the sediment.
Coast	The coast is defined as the part of the land adjoining or near the ocean. A coastline is properly a line on a map indicating the disposition of a coast, but the word is often used to refer to the coast itself. The adjective coastal describes something as being on, near to, or associated with a coast.
Shoreline	A shoreline is the fringe of land at the edge of a large body of water, such as an ocean, sea, or lake. A strict definition is the strip of land along a water body that is alternately exposed and covered by waves and tides.
Winter	Winter is one of the four seasons of temperate zones. Almost all English-language calendars, going by astronomy, state that winter begins on the winter solstice, and ends on the spring equinox. Calculated more by the weather, it begins and ends earlier and is the season with the shortest days and the lowest temperatures.
Sea level	Mean sea level is the average height of the sea, with reference to a suitable reference surface.
Aggradation	Aggradation in geology is the accumulation of sediment in rivers and nearby landforms. Aggradation occurs when sediment supply exceeds the ability of a river to transport the sediment.
Detritus	In biology, detritus is non-living particulate organic material. It typically includes the bodies of dead organisms or fragments of organisms or faecal material. Detritus is normally colonised by communities of microorganisms which act to decompose the material.
Pelagic	The pelagic is the part of the open sea or ocean that is not near the coast or sea floor. In contrast, the demersal zone comprises the water that is near to, and is significantly affected by, the coast or the sea floor.
Oolite	An oolite is a sedimentary rock formed from ooids, spherical grains composed of concentric layers.
Chalk	Chalk is a soft, white, porous sedimentary rock, a form of limestone composed of the mineral calcite. It forms under relatively deep marine conditions from the gradual accumulation of minute calcite plates shed from micro-organisms called coccolithophores. It is common to find

Go to **Cram101.com** for the Practice Tests for this Chapter.

flint nodules embedded in it.

Abundance	Abundance is an ecological concept referring to the relative representation of a species in a particular ecosystem. It is usually measured as the mean number of individuals found per sample.
Diagenesis	In geology and oceanography, diagenesis is any chemical, physical, or biological change undergone by a sediment after its initial deposition and during and after its lithification, exclusive of surface alteration, weathering and metamorphism. These changes happen at relatively low temperatures and pressures and result in changes to the rock face=symbol>¢s original mineralogy and texture
Delta	A delta is a landform where the mouth of a river flows into an ocean, sea, desert, estuary or lake. It builds up sediment outwards into the flat area which the river face=symbol>¢s flow encounters transported by the water and set down as the currents slow.
Salt marsh	A salt marsh is a type of marsh that is a transitional zone between land and salty or brackish water.
Lithification	Lithification is the process in which sediments compact under pressure, expel connate fluids, and gradually become solid rock.
Alluvial	An alluvial plain is a relatively flat and gently sloping landform found at the base of a range of hills or mountains, formed by the deposition of alluvial soil over a long period of time by one or more rivers coming from the mountains.
Tidal range	The tidal range is the vertical difference between the highest high tide and the lowest low tide. In other words, it is the difference in height between high and low tides.
Pelagic sediment	Pelagic sediment is an accumulate in the abyssal plain of the deep ocean, far away from terrestrial sources that provide terrigenous sediments; the latter are primarily limited to the continental shelf, and deposited by rivers.
Subsidence	In geology, engineering, and surveying, subsidence is the motion of a surface as it shifts downward relative to a datum such as sea-level. The opposite of subsidence is uplift, which results in an increase in elevation. In meteorology, subsidence refers to the downward movement of air.
Geostrophic	A geostrophic current results from the balance between gravitational forces and the Coriolis effect. The gravitational effect is controlled by the tilt of the sea surface and water density as controlled by horizontal changes in temperature and salinity.
Continental slope	The sea floor below the break is the continental slope. Below the slope is the continental rise, which finally merges into the deep ocean floor, the abyssal plain. As the continental shelf and the slope are part of the continental margin, both are covered in this article.
Convergent boundary	In plate tectonics, a convergent boundary is an actively deforming region where two tectonic plates or fragments of lithosphere move towards one another. When two plates move toward one another, they form either a subduction zone or a continental collision.
Blocks	Blocks in meteorology are large scale patterns in the atmospheric pressure field that are nearly stationary, effectively "blocking" or redirecting migratory cyclones. These blocks can remain in place for several days or even weeks, causing the areas affected by them to have the same kind of weather for an extended period of time.
Debris flow	Debris flow often refers to mudslides, mudflows, jökulhlaups, or debris avalanches. They consist primarily of geological material mixed with water. They may be generated when hillside colluvium or landslide material becomes rapidly saturated with water and flows into a channel.

Go to **Cram101.com** for the Practice Tests for this Chapter.
And, **NEVER** highlight a book again!

Benthos	Benthos are the organisms which live on, in, or near the seabed. Although the term derived from the Greek for "depths of the sea", the term is also used in freshwater biology to refer to organisms at the bottoms of freshwater bodies of water, such as lakes, rivers, and streams.
Invertebrate	Invertebrate is an English word that describes any animal without a spinal column.
Varve	A varve is an annual layer of sediment or sedimentary rock. The word Varve is derived from the Swedish word varv whose meanings and connotations include revolution, in layers, and circle.
Sedimentary rock	Sedimentary rock is one of the three main rock groups. Sedimentary rock covers 75% of the Earth's land area. Four basic processes are involved in the formation of a clastic sedimentary rock: weathering caused mainly by friction of waves, transportation where the sediment is carried along by a current, deposition and compaction where the sediment is squashed together to form a rock of this kind.
Chondrite	Chondrite refers to stony meteorites that have not been modified due to melting or differentiation of the parent body. They formed when various types of dust and small grains that were present in the early solar system accreted to form primitive asteroids.
Matter	Matter is the substance of which physical objects are composed. Matter can be solid, liquid, plasma or gas. It constitutes the observable universe.
Organic matter	Organic matter is matter that has come from a recently living organism; is capable of decay, or the product of decay; or is composed of organic compounds. The definition of organic matter varies upon the subject it is being used for.
Aerobic	An aerobic organism is an organism that has an oxygen based metabolism
Sulfide	The term sulfide refers to several types of chemical compounds containing sulfur in its lowest oxidation number of -2.
Canyon	A canyon is a deep valley between cliffs often carved from the landscape by a river. Most were formed by a process of long-time erosion from a plateau level. The cliffs form because harder rock strata that are resistant to erosion and weathering remain exposed on the valley walls.
Submarine canyon	A Submarine canyon is a steep-sided valley on the sea floor of the continental slope. They are formed by powerful turbidity currents, volcanic and earthquake activity. Many continue as submarine channels across continental rise areas and may extend for hundreds of kilometers.
Alvin	Alvin is a 16-ton, manned deep-ocean research submersible owned by the United States Navy and operated by the Woods Hole Oceanographic Institution in Woods Hole, Massachusetts. The three-person vessel allows for two scientists and one pilot to dive for up to nine hours at 4500 metersor 15,000 feet.
Mica	The mica group of sheet silicate minerals includes several closely related materials having highly perfect basal cleavage. All are monoclinic with a tendency towards pseudo-hexagonal crystals and are similar in chemical composition. The highly perfect cleavage, which is the most prominent characteristic of mica, is explained by the hexagonal sheet-like arrangement of its atoms.
Levee	A levee is a natural or artificial slope or wall, usually earthen and often parallels the course of a river.
Distributary channel	A distributary channel, is a stream that branches off and flows away from a main stream channel. They are a common feature of river deltas. The phenomenon is known as river bifurcation.

Go to **Cram101.com** for the Practice Tests for this Chapter.

Calcareous	Calcareous refers to a sediment, sedimentary rock, or soil type which is formed from or contains a high proportion of calcium carbonate in the form of calcite or aragonite.
Vesicle	In cell biology, a vesicle is a relatively small and enclosed compartment, separated from the cytosol by at least one lipid bilayer.
Slump	Slump is a form of mass wasting event that occurs when loosely consolidated materials or rock layers move a short distance down a slope. When the movement occurs in soil, there is often a distinctive rotational movement to the mass, that cuts vertically through bedding planes.
Stratovolcano	A stratovolcano, is a tall, conical volcano composed of many layers of hardened lava, tephra, and volcanic ash. These volcanoes are characterized by a steep profile and periodic, explosive eruptions. The lava that flows from them is viscous, and cools and hardens before spreading very far.
Scree	Scree is a term given to broken rock that appears at the bottom of crags, mountain cliffs or valley shoulders, forming a scree slope. The maximum inclination of such deposits corresponds to the angle of repose of the mean debris size.
Ice wedge	An ice wedge is a narrow mass of ice that can be 3 or 4 meters wide at ground surface and extend up to 10 meters downwards. During winter the ground gets very cold and the water in the ground freezes and expands. Then, as the temperature falls further, the soil and frozen ice acts as a solid and contracts as it gets colder, forming cracks.
Vertisol	Vertisol is a soil in which there is a high content of expansive clay known as montmorillonite that forms deep cracks in drier seasons or years. Alternate shrinking and swelling causes self-mulching, where the soil material consistently mixes itself.
Pāhoehoe	A pāhoehoe flow typically advances as a series of small lobes and toes that continually break out from a cooled crust. Also forms lava tubes where the minimal heat loss maintains low viscosity.
Basalt	Basalt is a common gray to black extrusive volcanic rock. It is usually fine-grained due to rapid cooling of lava on the Earth's surface. It may be porphyritic containing larger crystals in a fine matrix, or vesicular, or frothy scoria.
Lava	Lava is molten rock expelled by a volcano during an eruption. When first extruded from a volcanic vent, it is a liquid at temperatures from 700 °C to 1,200 °C.
Thrust	A thrust fault is a particular type of fault, or break in the fabric of the Earth face=symbol>¢s crust with resulting movement of each side against the other, in which a lower stratigraphic position is pushed up and over another. This is the result of compressional forces.
Pyroclastics	Pyroclastics are clastic rocks composed solely or primarily of volcanic materials.
Cleavage	Cleavage, in mineralogy, is the tendency of crystalline materials to split along definite planes, creating smooth surfaces.
Paleontologists	Paleontologists are people who study prehistoric life forms on Earth through the examination of plant and animal fossils. This includes the study of body fossils, tracks, burrows, cast-off parts, fossilised faeces, palynomorphs and chemical residues.
Spit	A spit is a deposition landform found off coasts. A spit is a type of bar or beach that develops where a re-entrant occurs, such as at a cove, bay, ria, or river mouth. A spit is formed by the movement of sediment along a shore by a process known as longshore drift. Where the direction of the shore turns inland the longshore current spreads out or dissipates. No longer able to carry the full load, much of the sediment is dropped. This causes a bar to build out from the shore, eventually becoming a spit.

Petrology	Petrology is a field of geology which focuses on the study of rocks and the conditions by which they form. There are three branches of petrology, corresponding to the three types of rocks: igneous, metamorphic, and sedimentary. Petrology utilizes the classical fields of mineralogy, petrography, optical mineralogy, and chemical analyses to describe the composition and texture of rocks.
Submarine fans	Submarine fans are underwater structures that look like deltas formed at the end of many large rivers, such as the Nile or Mississippi Rivers. They formed due to underwater currents. Close to land, a river deposits sediments onto the continental shelf. By then, the sediments are still suspended in the water, with some of the larger particles sinking to the floor of the continental shelf.
Sedimentology	Sedimentology encompasses the study of modern sediments and understanding the processes that deposit them. It also compares these observations to studies of ancient sedimentary rocks. Sedimentologists apply their understanding of modern processes to historically formed sedimentary rocks, allowing them to understand how they formed.
Miocene	The Miocene Epoch is a period of time that extends from about 23.03 to 5.332 million years before the present. As with other older geologic periods, the rock beds that define the start and end are well identified but the exact dates of the start and end of the period are uncertain.
Cenozoic	The Cenozoic Era meaning "new life", is the most recent of the three classic geological eras. It covers the 65.5 million years since the Cretaceous-Tertiary extinction event at the end of the Cretaceous that marked the demise of the last non-avian dinosaurs and the end of the Mesozoic Era. The Cenozoic era is ongoing.
Earth science	Earth science is an all-embracing term for the sciences related to the planet Earth. It is arguably a special case in planetary science, being the only known life-bearing planet. There are both reductionist and holistic approaches to Earth science. The major historic disciplines use physics, geology, geography, mathematics, chemistry, and biology to build a quantitative understanding of the principal areas or spheres of the Earth system.

Go to **Cram101.com** for the Practice Tests for this Chapter.
And, **NEVER** highlight a book again!

Quaternary	Quaternary refers to the period of Earth's history from about 2 million years ago to the present; also, the rocks and deposits of that age.
Primary structure	The primary structure of an unbranched biopolymer, such as a molecule of DNA, RNA or protein, is the specific nucleotide or peptide sequence from the beginning to the end of the molecule.
Paleosol	Paleosol refers to an ancient, buried soil whose composition may reflect a climate significantly different from the climate now prevalent in the area where the soil is found.
Soil	Soil refers to the top few meters of regolith, generally including some organic matter derived from plants.
Landform	Landform refers to any physical, recognizable form or feature on the earth's surface, having a characteristic shape and range in composition, and produced by natural causes.
Index fossil	The fossil of an organism known to have existed for a relatively short period of time, used to date the rock in which it is found, is called index fossil.
Fossil	A preserved remnant or impression of an organism that lived in the past is referred to as fossil.
Period	Period refers to a basic unit of the geologic time scale that is a subdivision of an era. Periods may be divided into smaller units called epochs.
Glacial drift	Glacial drift refers to the crushed rock and clays deposited by a melting glacier or ice sheet.
Interglacial	The term interglacial, such as the current era, is used to denote the absence of large-scale glaciation on a global scale — i.e., a non-Ice Age. They are, in general, shorter than glacial epochs.
Climate	Weather condition of an area including especially prevailing temperature and average daily/yearly rainfall over a long period of time is called climate.
Drift	Drift refers to a general term applied to all mineral material transported by a glacier and deposited directly by or from the ice, or by running water emanating from the glacier. Generally applies to Pleistocene glacial deposits.
Stratigraphy	Stratigraphy, a branch of geology, is basically the study of rock layers and layering. It is primarily used in the study of sedimentary and layered volcanic rocks.
Horizon	Horizon refers to levels within a soil profile that differ structurally and chemically. Generally divided into A, B, C, E, and 0 horizons.
Sediment	Sediment is any particulate matter that can be transported by fluid flow and which eventually is deposited as a layer of solid particles on the bed or bottom of a body of water or other liquid.
Geology	The scientific study of the Earth, its origins and evolution, the materials that make it up, and the processes that act on it is called geology.
Stage	Stage refers to the height of floodwaters in feet or meters above an established datum plane.
Conglomerate	Conglomerate refers to a clastic rock composed of particles more than 2 millimeters in diameter and marked by the roundness of its component grains and rock fragments.
Alluvial	Alluvial refers to pertaining to material or processes associated with transportation and or subaerial deposition by concentrated running water.
Clast	An individual grain or constituent of a rock is a clast.
Gravel	Rounded particles coarser than 2 mm in diameter are called gravel.

Floodplain	Floodplain refers to the flat land that surrounds a stream and becomes submerged when the stream overflows its banks.
Levee	A protective barrier built along the banks of a stream to prevent flooding is a levee.
Bioturbation	The process by which organisms rework existing sediments by burrowing through muds and sands is referred to as bioturbation.
Deposition	Any accumulation of material, by mechanical settling from water or air, chemical precipitation, evaporation from solution, etc is referred to as deposition.
Siltstone	A sedimentary rock consisting mostly of silt grains is called siltstone.
Crevasse	Crevasse refers to a crack in a glacier caused by rapid extension. Crevasses over 10 m deep would be healed by internal flow, but much deeper crevasses can be maintained by continued tension.
Channel	Channel refers to a feature on the surface of the planet Mars that very closely resembles certain types of stream channels on Earth.
Levees	Levees refer to embankments of sand or silt built by a stream along both banks of its channel. They are deposited during floods, when waters overflowing the stream banks are forced to deposit sediment.
Peat	Peat refers to soil type largely composed of partly decomposed organic material.
River	A major branch of a stream system is referred to as a river.
Aggradation	The process by which a stream's gradient steepens due to increased deposition of sediment is referred to as aggradation.
Valley	The entire area between the tops of the slopes on both sides of a stream is a valley.
Trough	Trough refers to the steep-walled, broad-floored shape considered diagnostic of former mountain glaciation. Often contrasted to the 'V' shape typical of mass wasting slopes feeding river systems.
Strata	Parallel layers of sedimentary rock are called strata.
Debris	Any unconsolidated material at Earth's surface is debris.
Debris flow	Debris flow refers to the rapid, downward mass movement of particles coarser than sand, often including boulders one meter or more in diameter, at a rate ranging from 2 to 40 kilometers per hour. Debris flows occur along fairly steep slopes.
Decomposition	Decomposition refers to the reduction of the body of a formerly living organism into simpler forms of matter.
Rocks	Aggregates of minerals or rock fragments are called rocks.
Rock	Rock refers to a naturally formed aggregate of usually inorganic materials from within the Earth.
Sorting	Sorting refers to the process by which a given transport medium separates out certain particles, as on the basis of size, shape, or density.
Talus	Talus refers to a pile of rock fragments lying at the bottom of the cliff or steep slope from which they have broken off.
Turbidity current	Turbidity current refers to a downslope movement of dense, sediment-laden water created when sand and mud on the continental shelf and slope are dislodged and thrown into suspension.
Evaporite	Inorganic chemical sediment that precipitates when the salty water in which it had dissolved evaporates is called evaporite.

Facies	Facies refers to a portion of a rock unit that possesses a distinctive set of characteristics that distinguishes it from other parts of the same unit.
Ripple	A sedimentary structure consisting of a very small dune of sand or silt whose long dimension is at right angles to the current is a ripple.
Delta	An alluvial fan having its apex at the mouth of a stream is a delta.
Tuff	Tuff refers to rock formed of pyroclastic material.
Cross-bedding	Cross-bedding refers to the structure in which relatively thin layers are inclined at an angle to the main bedding. Formed by currents of wind or water.
Accumulation	All processes that adds snow or ice to a glacier or to floating ice or snow cove are referred to as accumulation.
Bedding	The division of sediment or sedimentary rock into parallel layers that can be distinguished from each other by such features as chemical composition and grain size is bedding.
Eolian	Eolian refers to of, produced by, or carried by wind.
Dune	Dune refers to a usually asymmetrical mound or ridge of sand that has been transported and deposited by wind. Dunes form in both arid and humid climates.
Transverse dunes	Transverse dunes refer to a series of long ridges oriented at right angles to the prevailing wind; these dunes form where vegetation is sparse and sand is very plentiful.
Transverse dune	One of a series of dunes having an especially steep slip face and a gentle windward slope and standing perpendicular to the prevailing wind direction and parallel to each other. Transverse dunes typically form in arid and semi-arid regions with plentiful sand, stable wind direction, and scarce vegetation. A transverse dune may be as much as 100 kilometers long, 200 meters high, and 3 kilometers wide.
Parabolic dune	Parabolic dune refers to a horseshoe-shaped dune having a concave windward slope and a convex leeward slope. Parabolic dunes tend to form along sandy ocean and lake shores. They may also develop from transverse dunes through deflation.
Lamination	A thin layer in sedimentary rock is called lamination.
Barchan	Barchan refers to a crescent-shaped dune with limbs downwind.
Dip	The angle formed by the inclined plane of a geological structure and the horizontal plane of the Earth's surface is referred to as a dip.
Foreset bed	Foreset bed refers to an inclined bed deposited along the front of a delta.
Slump	Slump refers to a downward and outward slide occurring along a concave slip plane. The material that breaks off in such a slide.
Eolian dune	Eolian dune refers to a low mound, ridge, bank, or hill of loose, windblown, granular material, either bare or covered by vegetation, that is capable of movement from place to place but always maintaining its characteristic shape.
Pebble	A rock particle 2 to 64 mm in diameter is referred to as the pebble.
Glacier	A moving body of ice that forms on land from the accumulation and compaction of snow, and that flows downslope or outward due to gravity and the pressure of its own weight is a glacier.
Stream	Stream refers to a body of water found on the Earth's surface and confined to a narrow topographic depression, down which it flows and transports rock particles, sediment, and dissolved particles. Rivers, creeks, brooks, and runs are all streams.

Loess	A load of silt that is produced by the erosion of outwash and transported by wind is loess. Much loess found in the Mississippi Valley, China, and Europe is believed to have been deposited during the Pleistocene Epoch.
Till	Dominantly unsorted and unstratified drift, generally unconsolidated deposited directly by and underneath a glacier without subsequent reworking by meltwater, and consisting of a hetergeneous mixture of clay, silt, sand, gravel, stones, and boulders is called till.
Ablation till	A general term for loose, relatively permeable material deposited by the down wasting of nearly static glacier ice is referred to as ablation till.
Ablation	Ablation refers to all processes by which snow and ice are lost from a glacier, floating ice, or snow cover; or the amount which is melted. These processes include melting, evaporation, wind erosion, and calving.
Striations	Striations refer to multiple scratches or minute lines, generally parallel but occasionally cross-cutting, inscribed on a rock surface by a geologic agent. Common indicators of direction of glacier flow.
Striation	One of a group of usually parallel scratches engraved in bedrock by a glacier or other geological agent is striation.
Abrasion	A form of mechanical weathering that occurs when loose fragments or particles of rocks and minerals that are being transported, as by water or air, collide with each other or scrape the surfaces of stationary rocks is referred to as abrasion.
Rifting	Rifting refers to the tearing apart of a plate to form a depression in the Earth's crust and often eventually separating the plate into two or more smaller plates.
Moraine	Moraine is the general term for debris of all sorts originally transported by glaciers or ice sheets that have since melted away. Till is another word used to describe the sediments left by glaciers.
Detritus	Detritus is organic waste material from decomposing dead plants or animals.
Glacial lake	A lake that derives much or all of its water from the melting of glacier ice, fed by meltwater, and lying outside the glaciers margin is called the glacial lake.
Varve	Varve refers to a pair of sediment beds deposited by a lake on its floor, typically consisting of a thick, coarse, light-colored bed deposited in the summer and a thin, fine-grained, dark-colored bed deposited in the winter. Varves are most often found in lakes that freeze in the winter. The number and nature of varves on the bottom of a lake provide information about the lake's age and geologic events that affected the lake's development.
Fault	A fracture dividing a rock into two sections that have visibly moved relative to each other is a fault.
Fold	Fold refers to a bend that develops in an initially horizontal layer of rock, usually caused by plastic deformation. Folds occur most frequently in sedimentary rocks.
Silt	Silt refers to soil particles between the size of sand particles and clay particles, namely particles 0.002-0.2 mm in diameter.
Outwash	A load of sediment, consisting of sand and gravel that is deposited by meltwater in front of a glacier is called outwash.
Terminal moraine	An end moraine that marks the farthest advance of a glacier and usually has the form of a massive concentric ridge or complex of ridges, underlain by till and other types of drift is a terminal moraine.
Terminus	The outer margin of a glacier is called terminus.

Go to **Cram101.com** for the Practice Tests for this Chapter.

Tributary	Tributary refers to a stream that supplies water to a larger stream.
Kame	Kame refers to a low mound, knob, hummock, or short irregular ridge, composed of stratified sand and gravel deposited by a subglacial stream as a fan or delta at the margin of a melting glacier; or as a ponded deposit on the surface or at the margin of stagnant ice; or by a supraglacial stream in a low place or hole on the surface of the glacier.
Slab	Slab refers to a flat thinnish dressed stone.
Esker	A ridge of sediment that forms under a glacier's zone of ablation, made up of sand and gravel deposited by melt water is an esker. An esker may be less than 100 meters or more than 500 kilometers long, and may be anywhere from 3 to over 300 meters high.
Glaciation	A glaciation, often called an ice age, is a geological phenomenon in which massive ice sheets form in the Arctic and Antarctic and advance toward the equator.
Soil profile	A soil profile is a cross section through the soil which reveals its horizons (layers). Soil generally consists of visually and texturally distinct layers.
Creep	Creep refers to the slowest form of mass movement, measured in millimeters or centimeters per year and occurring on virtually all slopes.
Block	Angular chunk of solid rock ejected during an eruption is referred to as a block.
Pleistocene	A geologic epoch, characterized by alternating glacial and interglacial stages, that ended about 10,000 years ago, that lasted for 2 million years is called the pleistocene epoch.
Gravity	The force of attraction exerted by one body in the universe on another is gravity. Gravity is directly proportional to the product of the masses of the two attracted bodies. The force of attraction exerted by the Earth on bodies on or near its surface, tending to pull them toward the Earth's center.
Solifluction	A form of creep in which soil flows downslope at a rate of 0.5 to 15 centimeters per year is called solifluction. Solifluction occurs in relatively cold regions when the brief warmth of summer thaws only the upper meter or two of regolith, which becomes waterlogged because the underlying ground remains frozen and therefore the water cannot drain down into it.
Precursors	Observable phenomena that occur before an earthquake and indicate that an event is soon to occur are referred to as precursors.
Shear	Shear refers to stress that causes two adjacent parts of a body to slide past one another.
Lead	Lead is a chemical element in the periodic table that has the symbol Pb and atomic number 82. A soft, heavy, toxic and malleable poor metal, lead is bluish white when freshly cut but tarnishes to dull gray when exposed to air. Lead is used in building construction, lead-acid batteries, bullets and shot, and is part of solder, pewter, and fusible alloys.
Rotational slide	Rotational slide in mass wasting refers to a movement along a curved surface in which the upper part moves vertically downward while the lower part moves outward. Also called a slump.
Slide	The mass movement of a single, intact mass of rock, soil, or unconsolidated material along a weak plane, such as a fault, fracture, or bedding plane is a slide. A slide may involve as little as a minor displacement of soil or as much as the displacement of an entire mountainside.
Sedimentary rocks	Rocks formed by solidification of sediments formed and transported at the Earth's surface are referred to as sedimentary rocks.
Sedimentary rock	Sedimentary rock is one of the three main rock groups and is formed in three main ways—by the deposition of the weathered remains of other rocks; by the deposition of the results of biogenic activity; and by precipitation from solution.

Go to **Cram101.com** for the Practice Tests for this Chapter.

Go to **Cram101.com** for the Practice Tests for this Chapter.
And, **NEVER** highlight a book again!

Serpentine	An igneous rock rich in magnesium that forms soils toxic to many plants is called serpentine. Soils derived from serpentine are toxic to many plants due to their high mineral content, and the flora is generally very distinctive, with specialized, slow-growing species.
Foliation	The layering within metamorphic rocks is called foliation and it occurs when a strong compressive force is applied from one direction to a recrystallizing rock. This causes the platy or elongated crystals of minerals, such as mica and chlorite, to grow with their long axes perpendicular to the direction of the force.
Mineral	A naturally occurring, usually inorganic, solid consisting of either a single element or a compound, and having a definite chemical composition and a systematic internal arrangement of atoms is referred to as a mineral.
Ductile	Ductile refers to capable of being molded and bent under stress.
Cement	The solid material that precipitates in the pore space of sediments, binding the grains together to form solid rock is referred to as cement.
Salt	Salt is a term used for ionic compounds composed of positively charged cations and negatively charged anions, so that the product is neutral and without a net charge.
Earthflow	Earthflow refers to the downslope movement of water-saturated, clay-rich sediment. Mostly characteristic of humid regions.
Stratigraphic	The study of rock strata, especially of their distribution, deposition, and age is called stratigraphic.
Permafrost	Permanently frozen ground usually occurring in the tundra, a biome of Arctic regions is referred to as permafrost.
Melt	The liquid portion of magma excluding the solid crystals is called melt.
Dike	Dike refers to a discordant pluton that is substantially wider than it is thick. Dikes are often steeply inclined or nearly vertical.
Scale	The relationship between distance on a map and the distance on the terrain being represented by that map is a scale.
Mound	Of either earth or stone pebbles, generally covering a burial chamber or deposit is called mound.
Ash	Fine particles of pulverized rock blown from an explosion vent are called ash. Measuring less than 1/10 inch in diameter, ash may be either solid or molten when first erupted.
Carbon	Carbon is a chemical element in the periodic table that has the symbol C and atomic number 6. An abundant nonmetallic, tetravalent element, carbon has several allotropic forms.
Aluminum	Aluminium is the chemical element in the periodic table that has the symbol Al and atomic number 13. It is a silvery and ductile member of the poor metal group of chemical elements. Aluminium is found primarily as the ore bauxite and is remarkable for its resistance to corrosion (due to the phenomenon of passivation) and its light weight. Aluminium is used in many industries to make millions of different products and is very important to the world economy.
Leaching	Leaching is the process of extracting a substance from a solid by dissolving it in a liquid. In the chemical processing industry, leaching is known as extraction.
Quartz	Mineral with the formula SiO is referred to as quartz.
Iron	Iron is essential to all organisms, except for a few bacteria. It is mostly stably incorporated in the inside of metalloproteins, because in exposed or in free form it causes production of free radicals that are generally toxic to cells.

Go to **Cram101.com** for the Practice Tests for this Chapter.

Parent material	The rock material, the weathering and gradual breakdown of which is the source of the mineral portion of soil is referred to as parent material.
Clay minerals	Clay minerals refer to hydrous aluminum silicates that have a layered atomic structure. They are very fine grained and become plastic when wet. Most belong to one of three clay groups: kaolinite, illite, and smectite.
Clay mineral	A hydrous aluminum-silicate that occurs as a platy grain of microscopic size with a sheet silicate structure is clay mineral.
Oxide	One of several minerals containing negative oxygen ions bonded to one or more positive metallic ions are called oxide.
Humus	Humus is a complex organic substance resulting from the breakdown of plant material in a process called humification. This process can occur naturally in soil, or in the production of compost.
Bedrock	Bedrock refers to general term referring to the rock underlying other unconsolidated material, i.e. soil.
Bog	Bog refers to waterlogged, spongy ground, consisting of mosses containing acidic, decaying vegetation such as spaghnum, sedges, and heaths that may develop into peat.
Soil horizon	Soil horizon refers to a layer of soil formed at a characteristic depth and distinguished by its physical and chemical properties.
Volcanic ash	Volcanic ash refers to extremely small fragments, usually of glass, that forms when escaping gases force a fine spray of magma from a volcano.
B horizon	B horizon refers to a soil layer characterized by the accumulation of material leached downward from the A horizon above; also called zone of accumulation.
Fracture	Fracture refers to a crack or break in a rock. To break in random places instead of cleaving.
Plastic	Plastic refers to capable of being molded into any form, which is retained.
Evaporation	Evaporation refers to the change of state of water from the liquid to vapor phase. Requires the addition of 80 calories per cubic centimeter.
C horizon	A soil layer composed of incompletely weathered parent material is C horizon.
Carbonate	One of several minerals containing one central carbon atom with strong covalent bonds to three oxygen atoms and typically having ionic bonds to one or more positive ions is carbonate.
Opal	Opal refers to a mineraloid composed of silica and water.
Silicate mineral	Any one of numerous minerals that have the silicon-oxygen tetrahedron as their basic structure is called silicate mineral.
Dry climate	A climate in which yearly precipitation is less than the potential loss of water by evaporation is a dry climate.
Silicate	One of several rock-forming minerals that contain silicon, oxygen, and usually one or more other common elements is silicate.
Cation	Cation refers to a positively charged atom; an atom with fewer electrons than protons.
Cation exchange	A process in which positively charged minerals are made available to a plant when hydrogen ions in the soil displace mineral ions from the clay particles is referred to as cation exchange.
Correlation	The process of determining that two or more geographically distant rocks or rock strata

Go to **Cram101.com** for the Practice Tests for this Chapter.

	originated in the same time period is referred to as correlation.
Terrace	A flat, steplike surface that lines a stream above the floodplain, often paired one on each side of the stream, marking a former floodplain that existed at a higher level before regional uplift or an increase in discharge caused the stream to erode into it is called the terrace.
Erosion	Erosion is the displacement of solids (soil, mud, rock, and other particles) by the agents of wind, water, ice, movement in response to gravity, or living organisms.
Relief	The vertical difference between the summit of a mountain and the adjacent valley or plain is referred to as a relief.
Formation	A body of rock identified by lithic characteristics and stratigraphic position and is mappable at the earth's surface or traceable in the subsurface is a formation.
Sand dune	A mound of loose sand grains heaped up by the wind is a sand dune.
Ash fall	Ash fall refers to a 'rain' of airborne volcanic ash.
Fall	The situation in mass wasting that occurs when material free-falls or bounces down a cliff is called a fall.

Go to **Cram101.com** for the Practice Tests for this Chapter.

Stratigraphic	The study of rock strata, especially of their distribution, deposition, and age is called stratigraphic.
Ephemeral stream	Ephemeral stream refers to a stream that is usually dry because it carries water only in response to specific episodes of rainfall. Most desert streams are of this type.
Valley	The entire area between the tops of the slopes on both sides of a stream is a valley.
Stream	Stream refers to a body of water found on the Earth's surface and confined to a narrow topographic depression, down which it flows and transports rock particles, sediment, and dissolved particles. Rivers, creeks, brooks, and runs are all streams.
Dip	The angle formed by the inclined plane of a geological structure and the horizontal plane of the Earth's surface is referred to as a dip.
Outcrop	Any place where bedrock is visible on the surface of the Earth is referred to as outcrop.
Fault	A fracture dividing a rock into two sections that have visibly moved relative to each other is a fault.
Fold	Fold refers to a bend that develops in an initially horizontal layer of rock, usually caused by plastic deformation. Folds occur most frequently in sedimentary rocks.
Column	A feature found in caves that is formed when a stalactite and stalagmite join is referred to as a column.
Scale	The relationship between distance on a map and the distance on the terrain being represented by that map is a scale.
Rock	Rock refers to a naturally formed aggregate of usually inorganic materials from within the Earth.
Rocks	Aggregates of minerals or rock fragments are called rocks.
Formation	A body of rock identified by lithic characteristics and stratigraphic position and is mappable at the earth's surface or traceable in the subsurface is a formation.
Primary structure	The primary structure of an unbranched biopolymer, such as a molecule of DNA, RNA or protein, is the specific nucleotide or peptide sequence from the beginning to the end of the molecule.
Strata	Parallel layers of sedimentary rock are called strata.
Bedding	The division of sediment or sedimentary rock into parallel layers that can be distinguished from each other by such features as chemical composition and grain size is bedding.
Shale	Shale refers to a sedimentary rock composed of detrital sediment particles less than 0.004 millimeters in diameter. Shale tends to be red, brown, black, or gray, and usually originate in relatively still waters.
Soil	Soil refers to the top few meters of regolith, generally including some organic matter derived from plants.
Sandstone	Sandstone refers to a clastic rock composed of particles that range in diameter from 1/16 millimeter to 2 millimeters in diameter. Sandstones make up about 25% of all sedimentary rocks.
Limestone	Limestone is a sedimentary rock composed largely of the mineral calcite (calcium carbonate: $CaCO_3$). Limestone often contains variable amounts of silica in the form of chert or flint, as well as varying amounts of clay, silt and sand as disseminations, nodules, or layers within the rock.
Mudstone	A detrital sedimentary rock composed of clay-sized particles is mudstone.

Go to **Cram101.com** for the Practice Tests for this Chapter.

Porosity	The percentage of a soil, rock, or sediment's volume that is made up of pores is porosity.
Plastic deformation	A permanent strain that entails no rupture is called plastic deformation.
Deformation	General term for the processes of folding, faulting, shearing, compression, or extension of rocks as the result of various natural forces is called deformation.
Cleavage	The tendency of certain minerals to break along distinct planes in their crystal structures where the bonds are weakest is called cleavage. Cleavage is tested by striking or hammering a mineral, and is classified by the number of surfaces it produces and the angles between adjacent surfaces.
Plastic	Plastic refers to capable of being molded into any form, which is retained.
Hinge line	Hinge line refers to line about which a fold appears to be hinged. Line of maximum curvature of a folded surface.
Extension	Strain involving an increase in length is an extension. Extension can cause crustal thinning and faulting.
Strain	Strain refers to the change in the shape or volume of a rock that results from stress.
Folding	The processes by which crustal forces deform an area of crust so that layers of rock are pushed into folds are called folding.
Slump	Slump refers to a downward and outward slide occurring along a concave slip plane. The material that breaks off in such a slide.
Alignment	More or less straight row of standing stone is called an alignment.
Strike	The direction or trend of a bedding plane or fault, as it intersects the horizontal is referred to as a strike.
Angle of dip	A vertical angle measured downward from the horizontal plane to an inclined plane is an angle of dip.
Summit	The topographically highest hillslope position of a hillslope profile and exhibiting a nearly level surface is referred to as the summit.
Concretion	Hard, rounded mass that develops when a considerable amount of cementing material precipitates locally in a rock, often around an organic nucleus is called concretion.
Accessory	A mineral whose presence in a rock is not essential to the proper classification of the rock is an accessory.

Deformation	General term for the processes of folding, faulting, shearing, compression, or extension of rocks as the result of various natural forces is called deformation.
Sediment	Sediment is any particulate matter that can be transported by fluid flow and which eventually is deposited as a layer of solid particles on the bed or bottom of a body of water or other liquid.
Cleavage	The tendency of certain minerals to break along distinct planes in their crystal structures where the bonds are weakest is called cleavage. Cleavage is tested by striking or hammering a mineral, and is classified by the number of surfaces it produces and the angles between adjacent surfaces.
Fossil	A preserved remnant or impression of an organism that lived in the past is referred to as fossil.
Oolite	A small sphere of calcite precipitated from seawater is called oolite.
Strain	Strain refers to the change in the shape or volume of a rock that results from stress.
Rock	Rock refers to a naturally formed aggregate of usually inorganic materials from within the Earth.
Plate	Plate refers to rigid parts of the Earth's crust and part of the Earth's upper mantle that moves and adjoins each other along zones of seismic activity.
Extension	Strain involving an increase in length is an extension. Extension can cause crustal thinning and faulting.
Intrusion	Intrusion refers to the process of emplacement of magma in pre-existing rock. Also, the term refers to igneous rock mass so formed within the surrounding rock.
Shear	Shear refers to stress that causes two adjacent parts of a body to slide past one another.
Fault	A fracture dividing a rock into two sections that have visibly moved relative to each other is a fault.
Rocks	Aggregates of minerals or rock fragments are called rocks.
Lineation	A rock texture formed by alignment of rod-like features or grains is referred to as lineation.
Foliation	The layering within metamorphic rocks is called foliation and it occurs when a strong compressive force is applied from one direction to a recrystallizing rock. This causes the platy or elongated crystals of minerals, such as mica and chlorite, to grow with their long axes perpendicular to the direction of the force.
Schistose	The texture of a rock in which visible platy or needle-shaped minerals have grown essentially parallel to each other under the influence of directed pressure is schistose.
Fold	Fold refers to a bend that develops in an initially horizontal layer of rock, usually caused by plastic deformation. Folds occur most frequently in sedimentary rocks.
Vein	A sheetlike deposit of minerals precipitated in fractures or joints that are foreign to the host rock is called a vein.
Metamorphic rocks	Preexisting rocks that have been altered by heat, pressure, or chemically active fluids are metamorphic rocks.
Metamorphic rock	Metamorphic rock is the result of the transformation of a pre-existing rock type, the protolith, in a process called metamorphism, which means "change in form". The protolith is subjected to heat (greater than 150 degrees Celsius) and extreme pressure causing profound physical and/or chemical change.

Go to **Cram101.com** for the Practice Tests for this Chapter.

Porphyroblast	A large crystal surrounded by a much finer grained matrix of other minerals in a metamorphic rock is a porphyroblast.
Metamorphic	Metamorphic refers to the term from the Greek 'meta' and 'morph', commonly occurs to rocks which are subjected to increased heat and/or pressure. Also applies to the conversion of snow into glacial ice.
Inclusion	A piece of one rock unit contained within another is called inclusion. Inclusions are used in relative dating. The rock mass adjacent to the one containing the inclusion must have been there first in order to provide the fragment.
Crystal	Crystal is a solid in which the constituent atoms, molecules, or ions are packed in a regularly ordered, repeating pattern extending in all three spatial dimensions.
Foliated rock	Foliated rock refers to a metamorphic rock that displays foliation. Foliated rocks include slate, phyllite, schist, and gneiss.
Carbonate	One of several minerals containing one central carbon atom with strong covalent bonds to three oxygen atoms and typically having ionic bonds to one or more positive ions is carbonate.
Quartz	Mineral with the formula Si0 is referred to as quartz.
Fracture	Fracture refers to a crack or break in a rock. To break in random places instead of cleaving.
Slaty	Describing a rock that splits easily along nearly flat and parallel planes is referred to as slaty.
Matrix	Matrix refers to the solid matter in which a fossil or crystal is embedded.
Scale	The relationship between distance on a map and the distance on the terrain being represented by that map is a scale.
Plutonic rock	An intrusive rock formed inside the Earth is referred to as a plutonic rock.
Conglomerate	Conglomerate refers to a clastic rock composed of particles more than 2 millimeters in diameter and marked by the roundness of its component grains and rock fragments.
Phosphorite	A chemical or biochemical sedimentary rock composed of calcium phosphate precipitated from phosphaterich seawater and formed diagenetically by the interaction between muddy or carbonate sediments and the phosphate-rich water is phosphorite.
Oxidation	Oxidation refers to the loss of electrons from a substance involved in a redox reaction; always accompanies reduction.
Limestone	Limestone is a sedimentary rock composed largely of the mineral calcite (calcium carbonate: $CaCO_3$). Limestone often contains variable amounts of silica in the form of chert or flint, as well as varying amounts of clay, silt and sand as disseminations, nodules, or layers within the rock.
Mudstone	A detrital sedimentary rock composed of clay-sized particles is mudstone.
Plutonic	Plutonic refers to igneous rocks formed beneath the surface of the Earth; typically with large crystals due to the slowness of cooling.
Lapilli	Lapilli refers to literally, 'little stones.' Round to angular rock fragments, measuring 1/10 inch to 2 1/2 inches in diameter, which may be ejected in either a solid or molten state.
Basalt	Basalt is a common gray to black volcanic rock. It is usually fine-grained due to rapid cooling of lava on the Earth's surface.
Slate	A fine-grained, foliated metamorphic rock that develops from shale and tends to break into thin, flat sheets is called slate.

Go to **Cram101.com** for the Practice Tests for this Chapter.

Go to **Cram101.com** for the Practice Tests for this Chapter.
And, **NEVER** highlight a book again!

Clast	An individual grain or constituent of a rock is a clast.
Lava	Magma that comes to the Earth's surface through a volcano or fissure is referred to as lava.
Tuff	Tuff refers to rock formed of pyroclastic material.
Intermediate	Intermediate refers to a descriptive term applied to igneous rocks that are transitional between basic and acidic with silica between 54% and 65%.
Block	Angular chunk of solid rock ejected during an eruption is referred to as a block.
Folding	The processes by which crustal forces deform an area of crust so that layers of rock are pushed into folds are called folding.
Hinge line	Hinge line refers to line about which a fold appears to be hinged. Line of maximum curvature of a folded surface.
Outcrop	Any place where bedrock is visible on the surface of the Earth is referred to as outcrop.
Element	A chemical element, often called simply element, is a chemical substance that cannot be divided or changed into other chemical substances by any ordinary chemical technique. An element is a class of substances that contain the same number of protons in all its atoms.
Structural geology	The scientific study of the geological processes that deform the Earth's crust and create mountains is structural geology.
Geology	The scientific study of the Earth, its origins and evolution, the materials that make it up, and the processes that act on it is called geology.
Strike	The direction or trend of a bedding plane or fault, as it intersects the horizontal is referred to as a strike.
Dip	The angle formed by the inclined plane of a geological structure and the horizontal plane of the Earth's surface is referred to as a dip.
Limb	Portion of a fold shared by an anticline and a syncline is referred to as a limb.
Axial plane	A plane containing all of the hinge lines of a fold is called the axial plane.
Competence	The ability of a given stream to carry sediment, measured as the diameter of the largest particle that the stream can transport is called competence.
Viscosity	A fluid's resistance to flow is called viscosity. Viscosity increases as temperatures decreases.
Lamination	A thin layer in sedimentary rock is called lamination.
Ductile	Ductile refers to capable of being molded and bent under stress.
Arch	Bridge of rock left above an opening eroded in a headland by waves is an arch.
Ductile material	Ductile material refers to a material that undergoes smooth and continuous plastic deformation under increasing force and does not spring back to its original shape when the deforming force is released.
Concretion	Hard, rounded mass that develops when a considerable amount of cementing material precipitates locally in a rock, often around an organic nucleus is called concretion.
Shale	Shale refers to a sedimentary rock composed of detrital sediment particles less than 0.004 millimeters in diameter. Shale tends to be red, brown, black, or gray, and usually originate in relatively still waters.
Anticline	A convex fold in a rock, the central part of which contains the oldest section of rock is referred to as the anticline.

Go to **Cram101.com** for the Practice Tests for this Chapter.

Go to **Cram101.com** for the Practice Tests for this Chapter.
And, **NEVER** highlight a book again!

Diapir	Diapir refers to bodies of rock or magma that ascends within Earth's interior because they are less dense than the surrounding rock.
Dome	Dome refers to a round or oval bulge on the Earth's surface, containing the oldest section of rock in its raised, central part.
Bedding	The division of sediment or sedimentary rock into parallel layers that can be distinguished from each other by such features as chemical composition and grain size is bedding.
Solution	Usually slow but effective process of weathering and erosion in which rocks are dissolved by water is a solution.
Period	Period refers to a basic unit of the geologic time scale that is a subdivision of an era. Periods may be divided into smaller units called epochs.
Numerical age	Age given in years or some other unit of time is the numerical age.
Melange	A body of rocks consisting of large blocks of different rocks jumbled together with little continuity of contacts is referred to as a melange.
Slip	The distance that one face of a fault is displaced relative to the other is referred to as a slip.
Lead	Lead is a chemical element in the periodic table that has the symbol Pb and atomic number 82. A soft, heavy, toxic and malleable poor metal, lead is bluish white when freshly cut but tarnishes to dull gray when exposed to air. Lead is used in building construction, lead-acid batteries, bullets and shot, and is part of solder, pewter, and fusible alloys.
Tabular	Describing a feature such as an igneous pluton having two dimensions that are much longer than the third is referred to as tabular.
Breccia	A clastic rock composed of particles more than 2 millimeters in diameter and marked by the angularity of its component grains and rock fragments is called breccia.
Cohesion	The tendency of the molecules of a substance to stick together is referred to as cohesion.
Mineral	A naturally occurring, usually inorganic, solid consisting of either a single element or a compound, and having a definite chemical composition and a systematic internal arrangement of atoms is referred to as a mineral.
Faulting	The processes by which crustal forces cause a rock formation to break and slip along a fault are called faulting.
Clastic	Being or pertaining to a sedimentary rock composed primarily from fragments of preexisting rocks or fossils is called clastic.
Dike	Dike refers to a discordant pluton that is substantially wider than it is thick. Dikes are often steeply inclined or nearly vertical.
Geologic map	Geologic map refers to a map representing the geology of a given area.
Relief	The vertical difference between the summit of a mountain and the adjacent valley or plain is referred to as a relief.
Groundwater	Water stored beneath the surface in open pore spaces and fractures in rock is called groundwater.
Normal fault	Normal fault refers to a dip-slip fault marked by a generally steep dip along which the hanging wall has moved downward relative to the footwall.
Monocline	A fold in rock connecting two vertically offset, horizontal sections of sedimentary rocks is called the monocline.

Go to **Cram101.com** for the Practice Tests for this Chapter.

Go to **Cram101.com** for the Practice Tests for this Chapter.
And, **NEVER** highlight a book again!

Fissures	Fissures refer to elongated fractures or cracks on the slopes of a volcano. Fissure eruptions typically produce liquid flows, but pyroclastics may also be ejected.
Fissure	A crack in rock along which there is a distinct separation is the fissure.
Echelon	Echelon refers to set of geologic features that are in an overlapping or a staggered arrangement. Each is relatively short, but collectively they form a linear zone in which the strike of the individual features is oblique to that of the zone as a whole.
Accessory	A mineral whose presence in a rock is not essential to the proper classification of the rock is an accessory.
Strike-slip fault	Strike-slip fault refers to a fault in which two sections of rock have moved horizontally in opposite directions, parallel to the line of the fracture that divided them. Strike-slip faults are caused by shearing stress.
Reverse fault	Reverse fault refers to a dip-slip fault marked by a hanging wall that has moved upward relative to the footwall. Reverse faults are often caused by the convergence of lithospheric plates.
Earthquake	A movement within the Earth's crust or mantle, caused by the sudden rupture or repositioning of underground rocks as they release stress is an earthquake.
Alluvial fan	A triangular deposit of sediment left by a stream that has lost velocity upon entering a broad, relatively flat valley is referred to as an alluvial fan.
Alluvial	Alluvial refers to pertaining to material or processes associated with transportation and or subaerial deposition by concentrated running water.
Alluvium	A deposit of sediment left by a stream on the stream's channel or floodplain is called alluvium.
Terrace	A flat, steplike surface that lines a stream above the floodplain, often paired one on each side of the stream, marking a former floodplain that existed at a higher level before regional uplift or an increase in discharge caused the stream to erode into it is called the terrace.
Stream	Stream refers to a body of water found on the Earth's surface and confined to a narrow topographic depression, down which it flows and transports rock particles, sediment, and dissolved particles. Rivers, creeks, brooks, and runs are all streams.
Mound	Of either earth or stone pebbles, generally covering a burial chamber or deposit is called mound.
Alignment	More or less straight row of standing stone is called an alignment.
Peat	Peat refers to soil type largely composed of partly decomposed organic material.
Primary structure	The primary structure of an unbranched biopolymer, such as a molecule of DNA, RNA or protein, is the specific nucleotide or peptide sequence from the beginning to the end of the molecule.
Strata	Parallel layers of sedimentary rock are called strata.
Hanging wall	The overlying surface of an inclined fault plane is called a hanging wall.
Country rock	Country rock refers to any rock that was older than and intruded by an igneous body.
Gravel	Rounded particles coarser than 2 mm in diameter are called gravel.
Detachment fault	Major fault in a mountain belt above which rocks have been intensely folded and faulted is referred to as detachment fault.
Horizon	Horizon refers to levels within a soil profile that differ structurally and chemically.

Go to **Cram101.com** for the Practice Tests for this Chapter.

Generally divided into A, B, C, E, and 0 horizons.

Recumbent fold A fold overturned to such an extent that the limbs are essentially horizontal is referred to as recumbent fold.

Stratigraphic The study of rock strata, especially of their distribution, deposition, and age is called stratigraphic.

Attenuation To dilute or spread out the waste material is attenuation.

Schistosity Schistosity refers to a type of foliation characteristic of coarser-grained metamorphic rocks. Such rocks have a parallel arrangement of platy minerals such as the micas.

Zonation The distribution of organisms in bands or regions corresponding to changes in ecological conditions along a continuum, for example, intertidal zonation and elevational zonation.

Allochthonous Allochthonous is something formed elsewhere than its present location.

Gravity The force of attraction exerted by one body in the universe on another is gravity. Gravity is directly proportional to the product of the masses of the two attracted bodies. The force of attraction exerted by the Earth on bodies on or near its surface, tending to pull them toward the Earth's center.

Bedrock Bedrock refers to general term referring to the rock underlying other unconsolidated material, i.e. soil.

Basin A round or oval depression in the Earth's surface, containing the youngest section of rock in its lowest, central part is a basin.

Slab Slab refers to a flat thinnish dressed stone.

Erosion Erosion is the displacement of solids (soil, mud, rock, and other particles) by the agents of wind, water, ice, movement in response to gravity, or living organisms.

Oxide One of several minerals containing negative oxygen ions bonded to one or more positive metallic ions are called oxide.

Iron Iron is essential to all organisms, except for a few bacteria. It is mostly stably incorporated in the inside of metalloproteins, because in exposed or in free form it causes production of free radicals that are generally toxic to cells.

Lithified To change into stone as in the transformation of loose sand to sandstone is referred to as lithified.

Formation A body of rock identified by lithic characteristics and stratigraphic position and is mappable at the earth's surface or traceable in the subsurface is a formation.

Heterogeneous A heterogeneous compound, mixture, or other such object is one that consists of many different items, which are often not easily sorted or separated, though they are clearly distinct.

Stratified Stratified, formed, arranged, or laid down in layers is called stratified. The term refers to geologic deposits.

Oceanic trench Deep steep-sided depression in the ocean floor caused by the subduction of oceanic crust beneath either other oceanic crust or continental crust is an oceanic trench.

Trench An elongated depression in the seafloor produced by bending of oceanic crust during subduction is a trench.

Overturned fold A fold in which both limbs dip in the same direction is an overturned fold.

Abrasion A form of mechanical weathering that occurs when loose fragments or particles of rocks and

Go to **Cram101.com** for the Practice Tests for this Chapter.
And, **NEVER** highlight a book again!

minerals that are being transported, as by water or air, collide with each other or scrape the surfaces of stationary rocks is referred to as abrasion.

Go to **Cram101.com** for the Practice Tests for this Chapter.

Go to **Cram101.com** for the Practice Tests for this Chapter.
And, **NEVER** highlight a book again!

Magma chamber	The subterranean cavity containing the gas-rich liquid magma, which feeds a volcano, is a magma chamber.
Zonation	The distribution of organisms in bands or regions corresponding to changes in ecological conditions along a continuum, for example, intertidal zonation and elevational zonation.
Sorting	Sorting refers to the process by which a given transport medium separates out certain particles, as on the basis of size, shape, or density.
Crystal	Crystal is a solid in which the constituent atoms, molecules, or ions are packed in a regularly ordered, repeating pattern extending in all three spatial dimensions.
Magma	Molten rock that forms naturally within the Earth is magma. Magma may be either a liquid or a fluid mixture of liquid, crystals, and dissolved gases.
Tuff	Tuff refers to rock formed of pyroclastic material.
Ash	Fine particles of pulverized rock blown from an explosion vent are called ash. Measuring less than 1/10 inch in diameter, ash may be either solid or molten when first erupted.
Correlation	The process of determining that two or more geographically distant rocks or rock strata originated in the same time period is referred to as correlation.
Strata	Parallel layers of sedimentary rock are called strata.
Sedimentary rocks	Rocks formed by solidification of sediments formed and transported at the Earth's surface are referred to as sedimentary rocks.
Sedimentary rock	Sedimentary rock is one of the three main rock groups and is formed in three main ways—by the deposition of the weathered remains of other rocks; by the deposition of the results of biogenic activity; and by precipitation from solution.
Volcanic rock	An extrusive igneous rock is referred to as volcanic rock.
Volcanism	The set of geological processes that result in the expulsion of lava, pyroclastics, and gases at the Earth's surface is referred to as volcanism.
Fossil	A preserved remnant or impression of an organism that lived in the past is referred to as fossil.
Rocks	Aggregates of minerals or rock fragments are called rocks.
Rock	Rock refers to a naturally formed aggregate of usually inorganic materials from within the Earth.
Tephra	Materials of all types and sizes that are erupted from a crater or volcanic vent and deposited from the air are called tephra.
Lava	Magma that comes to the Earth's surface through a volcano or fissure is referred to as lava.
Intrusive rocks	Igneous rocks that have 'intruded' into the crust, hence they are slowly cooled and generally have phaneritic texture are called intrusive rocks.
Intrusive rock	An igneous rock formed by the entrance of magma into preexisting rock is called intrusive rock.
Fault	A fracture dividing a rock into two sections that have visibly moved relative to each other is a fault.
Dip	The angle formed by the inclined plane of a geological structure and the horizontal plane of the Earth's surface is referred to as a dip.
Pyroclastic flow	A rapid, extremely hot, downward stream of pyroclastics, air, gases, and ash ejected from an erupting volcano is called a pyroclastic flow. A pyroclastic flow may be as hot as 800ºC or

more and may move at speeds exceeding 150 kilometers per hour.

Pyroclastic	Being or pertaining to rock fragments formed in a volcanic eruption is referred to as pyroclastic.
Polarity	The magnetic positive or negative character of a magnetic pole is called polarity.
Pahoehoe	Hawaiian term for lava with a smooth, billowy, or ropy surface is pahoehoe.
Stream	Stream refers to a body of water found on the Earth's surface and confined to a narrow topographic depression, down which it flows and transports rock particles, sediment, and dissolved particles. Rivers, creeks, brooks, and runs are all streams.
Fold	Fold refers to a bend that develops in an initially horizontal layer of rock, usually caused by plastic deformation. Folds occur most frequently in sedimentary rocks.
Viscosity	A fluid's resistance to flow is called viscosity. Viscosity increases as temperatures decreases.
Spines	Horn-like projections formed upon a lava dome are called spines.
Shear	Shear refers to stress that causes two adjacent parts of a body to slide past one another.
Slab	Slab refers to a flat thinnish dressed stone.
Vesicle	A small air pocket or cavity formed in volcanic rock during solidification is a vesicle.
Pipe	Pipe refers to a vertical conduit through the Earth's crust below a volcano, through which magmatic materials have passed. Commonly filled with volcanic breccia and fragments of older rock.
Plate	Plate refers to rigid parts of the Earth's crust and part of the Earth's upper mantle that moves and adjoins each other along zones of seismic activity.
Core	The innermost layer of the Earth, consisting primarily of pure metals such as iron and nickel is the core. The core is the densest layer of the Earth, and is divided into the outer core, which is believed to be liquid, and the inner core, which is believed to be solid.
Lava tube	A tunnel formed when the surface of a lava flow cools and solidifies while the still-molten interior flows through and drains away is a lava tube.
Channel	Channel refers to a feature on the surface of the planet Mars that very closely resembles certain types of stream channels on Earth.
Cavern	A naturally formed underground chamber or series of chambers most commonly produced by solution activity in limestone is called a cavern.
Pegmatite	A coarse-grained igneous rock with exceptionally large crystals, formed from a magma that contains a high proportion of water is referred to as pegmatite.
Olivine	A ferromagnesian mineral is olivine.
Zeolite	A class of silicate minerals containing water in cavities within the crystal structure and formed by metamorphism at very low temperatures and pressures is zeolite.
Basalt	Basalt is a common gray to black volcanic rock. It is usually fine-grained due to rapid cooling of lava on the Earth's surface.
Columnar joints	Columnar joints refer to a pattern of cracks that forms during cooling of molten rock to generate columns.
Plateau	An elevated area with relatively little internal relief is called a plateau.
Column	A feature found in caves that is formed when a stalactite and stalagmite join is referred to

as a column.

Striations	Striations refer to multiple scratches or minute lines, generally parallel but occasionally cross-cutting, inscribed on a rock surface by a geologic agent. Common indicators of direction of glacier flow.
Striation	One of a group of usually parallel scratches engraved in bedrock by a glacier or other geological agent is striation.
Breccia	A clastic rock composed of particles more than 2 millimeters in diameter and marked by the angularity of its component grains and rock fragments is called breccia.
Matrix	Matrix refers to the solid matter in which a fossil or crystal is embedded.
Hyaloclastite	A deposit formed by the flowing or intrusion of lava or magma into water, ice, or water-saturated sediment and its consequent granulation or shattering into small angular fragments is hyaloclastite.
Clastic	Being or pertaining to a sedimentary rock composed primarily from fragments of preexisting rocks or fossils is called clastic.
Glass	A non-crystaline rock that results from very rapid cooling of magma is glass.
Block	Angular chunk of solid rock ejected during an eruption is referred to as a block.
Cement	The solid material that precipitates in the pore space of sediments, binding the grains together to form solid rock is referred to as cement.
Clast	An individual grain or constituent of a rock is a clast.
Massive	An igneous pluton that is not tabular in shape is massive.
Pillow lava	Lava extruded beneath water characterized by pillow-type shapes is called pillow lava.
Limestone	Limestone is a sedimentary rock composed largely of the mineral calcite (calcium carbonate: $CaCO_3$). Limestone often contains variable amounts of silica in the form of chert or flint, as well as varying amounts of clay, silt and sand as disseminations, nodules, or layers within the rock.
Chert	A member of a group of sedimentary rocks that consist primarily of microscopic silica crystals is chert. Chert may be either organic or inorganic, but the most common forms are inorganic.
Fissure eruption	An eruption in which lava is extruded from narrow fractures or cracks in the crust is called a fissure eruption.
Eruption	Eruption refers to the process by which solid, liquid, and gaseous materials are ejected into the earth's atmosphere and onto the earth's surface by volcanic activity. Eruptions range from the quiet overflow of liquid rock to the tremendously violent expulsion of pyroclastics.
Fissure	A crack in rock along which there is a distinct separation is the fissure.
Obsidian	Very hard volcanic glass used for tools. It can be dated by measurement of thickness of its hydration layer on surface is referred to as obsidian.
Gravity	The force of attraction exerted by one body in the universe on another is gravity. Gravity is directly proportional to the product of the masses of the two attracted bodies. The force of attraction exerted by the Earth on bodies on or near its surface, tending to pull them toward the Earth's center.
Diapir	Diapir refers to bodies of rock or magma that ascends within Earth's interior because they are less dense than the surrounding rock.

Go to **Cram101.com** for the Practice Tests for this Chapter.

Groundmass	The matrix of smaller crystals within an igneous rock that has porphyritic texture is referred to as groundmass.
Phenocryst	A conspicuous, usually large, crystal embedded in porphyritic igneous rock is a phenocryst.
Inclusion	A piece of one rock unit contained within another is called inclusion. Inclusions are used in relative dating. The rock mass adjacent to the one containing the inclusion must have been there first in order to provide the fragment.
Hematite	Hematite refers to a type of iron oxide that has a brick-red color when powdered.
Fracture	Fracture refers to a crack or break in a rock. To break in random places instead of cleaving.
Vent	An opening in the Earth's surface through which lava, gases, and hot particles are expelled is a vent. Also called volcanic vent and volcano.
Opal	Opal refers to a mineraloid composed of silica and water.
Dome	Dome refers to a round or oval bulge on the Earth's surface, containing the oldest section of rock in its raised, central part.
Extension	Strain involving an increase in length is an extension. Extension can cause crustal thinning and faulting.
Crater	A steep-sided, usually circular depression formed by either explosion or collapse at a volcanic vent is a crater.
Stratification	The arrangement of sedimentary rocks is called stratification.
Topography	Topography refers to the set of physical features, such as mountains, valleys, and the shapes of landforms, that characterizes a given landscape.
Fall	The situation in mass wasting that occurs when material free-falls or bounces down a cliff is called a fall.
Magnitude	A numerical expression of the amount of energy released by an earthquake, determined by measuring earthquake waves on standardized recording instruments is called a magnitude. The number scale for magnitudes is logarithmic rather than arithmetic. Therefore, deflections on a seismograph for a magnitude 5 earthquake, for example, are 10 times greater than those for a magnitude 4 earthquake, 100 times greater than for a magnitude 3 earthquake, and so on.
Detritus	Detritus is organic waste material from decomposing dead plants or animals.
Ejecta	Material that is thrown out by a volcano, including pyroclastic material and lava bombs is referred to as ejecta.
Glassy	A term used to describe the texture of certain igneous rocks, such as obsidian, that contain no crystals is referred to as glassy.
Scoria	Scoria refers to a bomb-size pyroclast that is irregular in form and generally very vesicular. It is usually heavier, darker, and more crystalline than pumice.
Bomb	Bomb refers to fragment of molten or semi-molten rock, 2 1/2 inches to many feet in diameter, which is blown out during an eruption. Because of their plastic condition, bombs are often modified in shape during their flight or upon impact.
Melt	The liquid portion of magma excluding the solid crystals is called melt.
Felsic	Felsic refers to a term used to describe the amount of light-colored feldspar and silica minerals in an igneous rock.
Pumice	Light-colored, frothy volcanic rock, usually of dacite or rhyolite composition, formed by the expansion of gas in erupting lava is pumice. Commonly seen as lumps or fragments of pea-size

Go to **Cram101.com** for the Practice Tests for this Chapter.

and larger, but can also occur abundantly as ash-sized particles.

Shards	Ash fragments of broken volcanic glass that compose tuff are shards.
Lapilli	Lapilli refers to literally, 'little stones.' Round to angular rock fragments, measuring 1/10 inch to 2 1/2 inches in diameter, which may be ejected in either a solid or molten state.
Facies	Facies refers to a portion of a rock unit that possesses a distinctive set of characteristics that distinguishes it from other parts of the same unit.
Volcano	The solid structure created when lava, gases, and hot particles escape to the Earth's surface through vents is called a volcano. Volcanoes are usually conical. A volcano is 'active' when it is erupting or has erupted recently. Volcanoes that have not erupted recently but are considered likely to erupt in the future are said to be 'dormant.' A volcano that has not erupted for a long time and is not expected to erupt in the future is 'extinct'.
Dune	Dune refers to a usually asymmetrical mound or ridge of sand that has been transported and deposited by wind. Dunes form in both arid and humid climates.
Crystallization	Crystallization refers to the formation and growth of a crystalline solid from a liquid or
Distal	Material that is deposited farthest from the source is referred to as distal.
Lithic	Of or pertaining to stone is referred to as lithic.
Vapor	Vapor refers to water in the gaseous state.
Compaction	The diagenetic process by which the volume or thickness of sediment is reduced due to pressure from overlying layers of sediment is called compaction.
Silicate	One of several rock-forming minerals that contain silicon, oxygen, and usually one or more other common elements is silicate.
Iron	Iron is essential to all organisms, except for a few bacteria. It is mostly stably incorporated in the inside of metalloproteins, because in exposed or in free form it causes production of free radicals that are generally toxic to cells.
Weathering	The process by which exposure to atmospheric agents, such as air or moisture, causes rocks and minerals to break down is called weathering. This process takes place at or near the Earth's surface. Weathering entails little or no movement of the material that it loosens from the rocks and minerals.
Compound	An electrically neutral substance that consists of two or more elements combined in specific, constant proportions is a compound. A compound typically has physical characteristics different from those of its constituent elements.
Thrust fault	A reverse fault marked by a dip of 45° or less is called thrust fault.
Lava flow	Lava flow refers to an outpouring of lava onto the land surface from a vent or fissure. Also, a solidified tongue like or sheet-like body formed by outpouring lava.
Sediment	Sediment is any particulate matter that can be transported by fluid flow and which eventually is deposited as a layer of solid particles on the bed or bottom of a body of water or other liquid.
Soil	Soil refers to the top few meters of regolith, generally including some organic matter derived from plants.
Porosity	The percentage of a soil, rock, or sediment's volume that is made up of pores is porosity.
Lahar	A flow of pyroclastic material mixed with water is called a lahar. A lahar is often produced when a snow-capped volcano erupts and hot pyroclastics melt a large amount of snow or ice.

Snow	Distinct crystals of ice are called snow. Commonly accumulates with a density of 50 - 200 kg·m, although wind-abraded and packed snow may have a higher initial density.
Debris flow	Debris flow refers to the rapid, downward mass movement of particles coarser than sand, often including boulders one meter or more in diameter, at a rate ranging from 2 to 40 kilometers per hour. Debris flows occur along fairly steep slopes.
Avalanche	A large mass of material or mixtures of material falling or sliding rapidly under the force of gravity is an avalanche. Avalanches often are classified by their content, such as snow, ice, soil, or rock avalanches. A mixture of these materials is a debris avalanche.
Caldera	A vast depression at the top of a volcanic cone, formed when an eruption substantially empties the reservoir of magma beneath the cone's summit is a caldera. A caldera may be more than 15 kilometers in diameter and more than 1000 meters deep.
Debris	Any unconsolidated material at Earth's surface is debris.
Bedding	The division of sediment or sedimentary rock into parallel layers that can be distinguished from each other by such features as chemical composition and grain size is bedding.
Talus	Talus refers to a pile of rock fragments lying at the bottom of the cliff or steep slope from which they have broken off.
Intercalation	Intercalation is a term used in host-guest chemistry for the reversible inclusion of a molecule (or group) between two other molecules (or groups). The host molecules usually comprise some form of periodic network.
Dike	Dike refers to a discordant pluton that is substantially wider than it is thick. Dikes are often steeply inclined or nearly vertical.
Elevation	Elevation refers to the altitude, or vertical distance, above or below sea level.
Episode	An episode is a volcanic event that is distinguished by its duration or style.
Subsidence	Subsidence is a term used in geology, engineering and surveying to denote the motion of a surface (usually, the earth's surface) downwards relative to a datum such as sea-level.
Columnar jointing	Columnar jointing refers to vertically oriented polygonal columns in a solid lava flow or shallow intrusion formed by contraction upon cooling. They are usually formed in basaltic lavas.
Sill	Sill refers to a concordant pluton that is substantially wider than it is thick. Sills form within a few kilometers of the Earth's surface.
Mafic	Mafic refers to a term used to describe the amount of dark-colored iron and magnesium minerals in an igneous rock.
Unconformity	A boundary separating two or more rocks of markedly different ages, marking a gap in the geologic record is referred to as unconformity.
Volcanic neck	A massive pillar of rock more resistant to erosion than the lavas and pyroclastic rocks of a volcanic cone is called a volcanic neck.
Ordovician	The Ordovician period is the second of the six (seven in North America) periods of the Paleozoic era. It follows the Cambrian period and is followed by the Silurian period.
Formation	A body of rock identified by lithic characteristics and stratigraphic position and is mappable at the earth's surface or traceable in the subsurface is a formation.
Ash flow	Ash flow refers to a turbulent mixture of gas and rock fragments, most of which are ash-sized particles, ejected violently from a crater or fissure. The mass of pyroclastics is normally of very high temperature and moves rapidly down the slopes or even along a level surface.

Resource	A mineral or fuel deposit, known or not yet discovered, that may be or become available for human exploitation is called a resource.
Geology	The scientific study of the Earth, its origins and evolution, the materials that make it up, and the processes that act on it is called geology.
Basaltic composition	A compositional group of igneous rocks indicating that the rock contains substantial dark silicate minerals and calcium-rich plagioclase feldspar is a basaltic composition.
Fluidization	The process whereby granular solids under high gas pressures become fluid-like and flow downslope or can be pumped is referred to as fluidization.
Diatreme	A breccia filled volcanic pipe that was formed by a gaseous explosion is a diatreme.
Mountain	A large mass of rock projecting above surrounding terrain is called a mountain.

Go to **Cram101.com** for the Practice Tests for this Chapter.

Hornblende	Hornblende refers to common amphibole frequently found in igneous and metamorphic rocks.
Polymorph	Polymorph refers to a mineral that is identical to another mineral in chemical composition but differs from it in crystal structure.
Mineral	A naturally occurring, usually inorganic, solid consisting of either a single element or a compound, and having a definite chemical composition and a systematic internal arrangement of atoms is referred to as a mineral.
Augite	Augite refers to a mineral of the pyroxene group found in mafic igneous rocks.
Country rock	Country rock refers to any rock that was older than and intruded by an igneous body.
Pluton	An intrusive rock, as distinguished from the preexisting country rock that surrounds it is called pluton.
Rocks	Aggregates of minerals or rock fragments are called rocks.
Rock	Rock refers to a naturally formed aggregate of usually inorganic materials from within the Earth.
Plutonic rock	An intrusive rock formed inside the Earth is referred to as a plutonic rock.
Inclusion	A piece of one rock unit contained within another is called inclusion. Inclusions are used in relative dating. The rock mass adjacent to the one containing the inclusion must have been there first in order to provide the fragment.
Plutonic	Plutonic refers to igneous rocks formed beneath the surface of the Earth; typically with large crystals due to the slowness of cooling.
Matrix	Matrix refers to the solid matter in which a fossil or crystal is embedded.
Schistose	The texture of a rock in which visible platy or needle-shaped minerals have grown essentially parallel to each other under the influence of directed pressure is schistose.
Evolution	In biology, evolution is the process by which novel traits arise in populations and are passed on from generation to generation. Its action over large stretches of time explains the origin of new species and ultimately the vast diversity of the biological world.
Ductile	Ductile refers to capable of being molded and bent under stress.
Shear	Shear refers to stress that causes two adjacent parts of a body to slide past one another.
Dike	Dike refers to a discordant pluton that is substantially wider than it is thick. Dikes are often steeply inclined or nearly vertical.
Crystal	Crystal is a solid in which the constituent atoms, molecules, or ions are packed in a regularly ordered, repeating pattern extending in all three spatial dimensions.
Magma	Molten rock that forms naturally within the Earth is magma. Magma may be either a liquid or a fluid mixture of liquid, crystals, and dissolved gases.
Melt	The liquid portion of magma excluding the solid crystals is called melt.
Discordant	A term used to describe plutons that cut across existing rock structures, such as bedding planes is called discordant.
Gneiss	Gneiss refers to a coarse-grained, foliated metamorphic rock marked by bands of light-colored minerals such as quartz and feldspar that alternate with bands of dark-colored minerals. This alternation develops through metamorphic differentiation.
Dome	Dome refers to a round or oval bulge on the Earth's surface, containing the oldest section of rock in its raised, central part.

Go to **Cram101.com** for the Practice Tests for this Chapter.
And, **NEVER** highlight a book again!

Viscosity	A fluid's resistance to flow is called viscosity. Viscosity increases as temperatures decreases.
Halogen	Halogen is a chemical series. They are the elements in Group 17 (old-style: VII or VIIA) of the periodic table: fluorine (F), chlorine (Cl), bromine (Br), iodine (I), astatine (At) and the as yet undiscovered ununseptium (Uus). The term halogen was coined to mean elements which produce salt in union with a metal.
Intrusion	Intrusion refers to the process of emplacement of magma in pre-existing rock. Also, the term refers to igneous rock mass so formed within the surrounding rock.
Rubidium	Rubidium is a chemical element in the periodic table that has the symbol Rb and atomic number 37. Rb is a soft, silvery-white metallic element of the alkali metal group. Rb-87, a naturally occurring isotope, is (slightly) radioactive. Rubidium is highly reactive, with properties similar to other elements in group 1, like igniting spontaneously in air.
Potassium	Potassium is a chemical element in the periodic table. It has the symbol K (L. kalium) and atomic number 19. Potassium is a soft silvery-white metallic alkali metal that occurs naturally bound to other elements in seawater and many minerals.
Contact metamorphism	Contact metamorphism is the name given to the changes that take place when magma is injected into the surrounding solid rock (country rock). The changes that occur are greatest wherever the magma comes into contact with the rock because the temperatures are highest at this boundary and decrease with distance from it.
Volcanic rock	An extrusive igneous rock is referred to as volcanic rock.
Metamorphism	The process by which conditions within the Earth, below the zone of diagenesis, alter the mineral content, chemical composition, and structure of solid rock without melting it is called metamorphism. Igneous, sedimentary, and metamorphic rocks may all undergo metamorphism.
Stratigraphy	Stratigraphy, a branch of geology, is basically the study of rock layers and layering. It is primarily used in the study of sedimentary and layered volcanic rocks.
Subsidence	Subsidence is a term used in geology, engineering and surveying to denote the motion of a surface (usually, the earth's surface) downwards relative to a datum such as sea-level.
Erosion	Erosion is the displacement of solids (soil, mud, rock, and other particles) by the agents of wind, water, ice, movement in response to gravity, or living organisms.
Stoping	Stoping refers to upward movement of a body of magma by fracturing of overlying country rock. Magma engulfs the blocks of fractured country rock as it moves upward.
Block	Angular chunk of solid rock ejected during an eruption is referred to as a block.
Crystallization	Crystallization refers to the formation and growth of a crystalline solid from a liquid or
Stage	Stage refers to the height of floodwaters in feet or meters above an established datum plane.
Groundmass	The matrix of smaller crystals within an igneous rock that has porphyritic texture is referred to as groundmass.
Phenocryst	A conspicuous, usually large, crystal embedded in porphyritic igneous rock is a phenocryst.
Olivine	A ferromagnesian mineral is olivine.
Biotite	Biotite refers to iron/magnesium-bearing mica.
Facies	Facies refers to a portion of a rock unit that possesses a distinctive set of characteristics that distinguishes it from other parts of the same unit.
Gabbro	Gabbro refers to any of a group of dark, dense, phaneritic, intrusive rocks that are the

plutonic equivalent to basalt.

Quartz	Mineral with the formula Si0 is referred to as quartz.
Plate	Plate refers to rigid parts of the Earth's crust and part of the Earth's upper mantle that moves and adjoins each other along zones of seismic activity.
Cleavage	The tendency of certain minerals to break along distinct planes in their crystal structures where the bonds are weakest is called cleavage. Cleavage is tested by striking or hammering a mineral, and is classified by the number of surfaces it produces and the angles between adjacent surfaces.
Gradient	The vertical drop in a stream's elevation over a given horizontal distance, expressed as an angle is referred to as a gradient.
Core	The innermost layer of the Earth, consisting primarily of pure metals such as iron and nickel is the core. The core is the densest layer of the Earth, and is divided into the outer core, which is believed to be liquid, and the inner core, which is believed to be solid.
Accumulation	All processes that adds snow or ice to a glacier or to floating ice or snow cove are referred to as accumulation.
Substrate	A substrate is a molecule which is acted upon by an enzyme. Each enzyme recognizes only the specific substrate of the reaction it catalyzes. A surface in or on which an organism lives.
Tabular	Describing a feature such as an igneous pluton having two dimensions that are much longer than the third is referred to as tabular.
Xenoliths	Xenoliths refer to a foreign inclusion in an igneous rock.
Formation	A body of rock identified by lithic characteristics and stratigraphic position and is mappable at the earth's surface or traceable in the subsurface is a formation.
Streak	Streak refers to the color of a mineral in its powdered form. This color is usually determined by rubbing the mineral against an unglazed porcelain slab and observing the mark made by it on the slab.
Igneous	Igneous rocks are formed when molten rock (magma) cools and solidifies, with or without crystallization, either below the surface as intrusive (plutonic) rocks or on the surface as extrusive (volcanic) rocks. This magma can be derived from either the Earth's mantle or pre-existing rocks made molten by extreme temperature and pressure changes.
Correlation	The process of determining that two or more geographically distant rocks or rock strata originated in the same time period is referred to as correlation.
Fracture	Fracture refers to a crack or break in a rock. To break in random places instead of cleaving.
Partial melting	The incomplete melting of a rock composed of minerals with differing melting points is called partial melting When partial melting occurs, the minerals with higher melting points remain solid while the minerals whose melting points have been reached turn to magma.
Mafic	Mafic refers to a term used to describe the amount of dark-colored iron and magnesium minerals in an igneous rock.
Deformation	General term for the processes of folding, faulting, shearing, compression, or extension of rocks as the result of various natural forces is called deformation.
Horizon	Horizon refers to levels within a soil profile that differ structurally and chemically. Generally divided into A, B, C, E, and 0 horizons.
Foliation	The layering within metamorphic rocks is called foliation and it occurs when a strong compressive force is applied from one direction to a recrystallizing rock. This causes the

Go to **Cram101.com** for the Practice Tests for this Chapter.
And, **NEVER** highlight a book again!

platy or elongated crystals of minerals, such as mica and chlorite, to grow with their long axes perpendicular to the direction of the force.

Extension	Strain involving an increase in length is an extension. Extension can cause crustal thinning and faulting.
Sill	Sill refers to a concordant pluton that is substantially wider than it is thick. Sills form within a few kilometers of the Earth's surface.
Fault	A fracture dividing a rock into two sections that have visibly moved relative to each other is a fault.
Ion	Ion refers to an atom or molecule that has gained or lost one or more electrons, thus acquiring an electrical charge.
Assimilation	In igneous activity, the process of incorporating country rock into a magma body is called assimilation.
Pegmatite	A coarse-grained igneous rock with exceptionally large crystals, formed from a magma that contains a high proportion of water is referred to as pegmatite.
Cross-bedding	Cross-bedding refers to the structure in which relatively thin layers are inclined at an angle to the main bedding. Formed by currents of wind or water.
Migmatite	Migmatite refers to a rock that incorporates both metamorphic and igneous materials.
Dissolution	A form of chemical weathering in which water molecules, sometimes in combination with acid or another compound in the environment, attract and remove oppositely charged ions or ion groups from a mineral or rock is a dissolution.
Volatiles	Gaseous components of magma dissolved in the melt are called volatiles. Volatiles will readily vaporize at surface pressures.
Peridotite	An igneous rock composed primarily of the iron-magnesium silicate olivine and having a silica content of less than 40% is peridotite.
Intermediate	Intermediate refers to a descriptive term applied to igneous rocks that are transitional between basic and acidic with silica between 54% and 65%.
Granite	A pink-colored, felsic, plutonic rock that contains potassium and usually sodium feldspars, and has quartz content of about 10% is granite. Granite is commonly found on continents but virtually absent from the ocean basins.
Vesicle	A small air pocket or cavity formed in volcanic rock during solidification is a vesicle.
Carbonate	One of several minerals containing one central carbon atom with strong covalent bonds to three oxygen atoms and typically having ionic bonds to one or more positive ions is carbonate.
Mafic rock	Silica-deficient igneous rock with a relatively high content of magnesium, iron, and calcium is called mafic rock.
Conglomerate	Conglomerate refers to a clastic rock composed of particles more than 2 millimeters in diameter and marked by the roundness of its component grains and rock fragments.
Deposition	Any accumulation of material, by mechanical settling from water or air, chemical precipitation, evaporation from solution, etc is referred to as deposition.
Faceted	A rock fragment with one or more flat surfaces caused by erosive action is called faceted.
Breccia	A clastic rock composed of particles more than 2 millimeters in diameter and marked by the angularity of its component grains and rock fragments is called breccia.

Go to **Cram101.com** for the Practice Tests for this Chapter.

Pebble	A rock particle 2 to 64 mm in diameter is referred to as the pebble.
Magnetite	Magnetite is a ferrimagnetic mineral form of iron(II,III) oxide, with chemical formula Fe_3O_4, one of several iron oxides and a member of the spinel group.
Hematite	Hematite refers to a type of iron oxide that has a brick-red color when powdered.
Magmatic	Magmatic refers to pertaining to magma.
Sulfide	One of the minerals that is abundant in the hot water that seeps through hydrothermal vents is sulfide.
Lineation	A rock texture formed by alignment of rod-like features or grains is referred to as lineation.
Fold	Fold refers to a bend that develops in an initially horizontal layer of rock, usually caused by plastic deformation. Folds occur most frequently in sedimentary rocks.
Plastic	Plastic refers to capable of being molded into any form, which is retained.
Strain	Strain refers to the change in the shape or volume of a rock that results from stress.
Strike	The direction or trend of a bedding plane or fault, as it intersects the horizontal is referred to as a strike.
Vein	A sheetlike deposit of minerals precipitated in fractures or joints that are foreign to the host rock is called a vein.
Groundwater	Water stored beneath the surface in open pore spaces and fractures in rock is called groundwater.
Porphyry	An igneous rock with a porphyritic texture is referred to as porphyry.
Stable	Describing a mineral that will not react with or convert to a new mineral or substance, given enough time is referred to as stable.
Geothermal	Geothermal is the naturally hot interior of Earth. The heat is maintained by naturally occurring nuclear reactions in Earth's interior.
Fracturing	Cracking or rupturing of a body under stress is referred to as fracturing.
Aureole	Zone of contact metamorphism adjacent to a pluton is called the aureole.
Ore deposit	The same as a mineral reserve except that it refers only to a metal-bearing deposit is referred to as ore deposit.
Ore	A mineral deposit that can be mined for a profit is called ore.
Potassium-argon dating	A form of isotope dating that relies on the extremely long half-life of radioactive isotopes of potassium, which decay into isotopes of argon, to determine the age of rocks in which argon is present. Potassium-argon dating is used for rocks between 100,000 and 4 billion years old.
Mineral resource	All discovered and undiscovered deposits of a useful mineral that can be extracted now or at some time in the future are called mineral resource.
Geophysics	The branch of geology that studies the physics of the Earth, using the physical principles underlying such phenomena as seismic waves, heat flow, gravity, and magnetism to investigate planetary properties is called geophysics.
Ultramafic	Ultramafic refers to igneous rocks made mostly of the mafic minerals hypersthene, augite, and/or olivine.
Resource	A mineral or fuel deposit, known or not yet discovered, that may be or become available for

Go to **Cram101.com** for the Practice Tests for this Chapter.

Go to **Cram101.com** for the Practice Tests for this Chapter.
And, **NEVER** highlight a book again!

human exploitation is called a resource.

Geology	The scientific study of the Earth, its origins and evolution, the materials that make it up, and the processes that act on it is called geology.
Mantle	The middle layer of the Earth, lying just below the crust and consisting of relatively dense rocks is called the mantle. The mantle is divided into two sections, the upper mantle and the lower mantle; the lower mantle has greater density than the upper mantle.
Crust	A crust is the outer layer of a planet, part of its lithosphere. Planetary crust is generally composed of a less dense material than that of its deeper layers. The crust of the Earth is composed mainly of basalt and granite.
Porphyry copper deposit	A crystallized rock, typically porphyritic, having hairline fractures that contain copper and other metals is called porphyry copper deposit.
Porphyry copper	A copper deposit, usually of low grade, in which the copper-bearing minerals occur in disseminated grains and/or in veinlets through a large volume of rock, is called porphyry copper.
Ultramafic rock	Rock composed entirely or almost entirely of ferromagnesian minerals is referred to as ultramafic rock.
Copper	Copper is a chemical element in the periodic table that has the symbol Cu (L.: Cuprum) and atomic number 29. It is a ductile metal with excellent electrical conductivity, and finds extensive use as a building material, as an electrical conductor, and as a component of various alloys.

Go to **Cram101.com** for the Practice Tests for this Chapter.
And, **NEVER** highlight a book again!

Metamorphic rocks	Preexisting rocks that have been altered by heat, pressure, or chemically active fluids are metamorphic rocks.
Metamorphic rock	Metamorphic rock is the result of the transformation of a pre-existing rock type, the protolith, in a process called metamorphism, which means "change in form". The protolith is subjected to heat (greater than 150 degrees Celsius) and extreme pressure causing profound physical and/or chemical change.
Metamorphism	The process by which conditions within the Earth, below the zone of diagenesis, alter the mineral content, chemical composition, and structure of solid rock without melting it is called metamorphism. Igneous, sedimentary, and metamorphic rocks may all undergo metamorphism.
Metamorphic	Metamorphic refers to the term from the Greek 'meta' and 'morph', commonly occurs to rocks which are subjected to increased heat and/or pressure. Also applies to the conversion of snow into glacial ice.
Rocks	Aggregates of minerals or rock fragments are called rocks.
Rock	Rock refers to a naturally formed aggregate of usually inorganic materials from within the Earth.
Mineral	A naturally occurring, usually inorganic, solid consisting of either a single element or a compound, and having a definite chemical composition and a systematic internal arrangement of atoms is referred to as a mineral.
Formation	A body of rock identified by lithic characteristics and stratigraphic position and is mappable at the earth's surface or traceable in the subsurface is a formation.
Stratigraphic	The study of rock strata, especially of their distribution, deposition, and age is called stratigraphic.
Aluminum	Aluminium is the chemical element in the periodic table that has the symbol Al and atomic number 13. It is a silvery and ductile member of the poor metal group of chemical elements. Aluminium is found primarily as the ore bauxite and is remarkable for its resistance to corrosion (due to the phenomenon of passivation) and its light weight. Aluminium is used in many industries to make millions of different products and is very important to the world economy.
Silicate	One of several rock-forming minerals that contain silicon, oxygen, and usually one or more other common elements is silicate.
Quartz	Mineral with the formula Si0 is referred to as quartz.
Silt	Silt refers to soil particles between the size of sand particles and clay particles, namely particles 0.002-0.2 mm in diameter.
Cross-cutting	Cross-cutting refers to a principle of relative dating. A rock or fault is younger than any rock through which it cuts.
Intrusion	Intrusion refers to the process of emplacement of magma in pre-existing rock. Also, the term refers to igneous rock mass so formed within the surrounding rock.
Igneous	Igneous rocks are formed when molten rock (magma) cools and solidifies, with or without crystallization, either below the surface as intrusive (plutonic) rocks or on the surface as extrusive (volcanic) rocks. This magma can be derived from either the Earth's mantle or pre-existing rocks made molten by extreme temperature and pressure changes.
Igneous rock	A igneous rock is formed when molten rock (magma) cools and solidifies, with or without crystallization, either below the surface as intrusive (plutonic) rocks or on the surface as extrusive (volcanic) rocks. This magma can be derived from either the Earth's mantle or pre-

Go to **Cram101.com** for the Practice Tests for this Chapter.

Go to **Cram101.com** for the Practice Tests for this Chapter.
And, **NEVER** highlight a book again!

existing rocks made molten by extreme temperature and pressure changes.

Sandstone	Sandstone refers to a clastic rock composed of particles that range in diameter from 1/16 millimeter to 2 millimeters in diameter. Sandstones make up about 25% of all sedimentary rocks.
Peridotite	An igneous rock composed primarily of the iron-magnesium silicate olivine and having a silica content of less than 40% is peridotite.
Olivine	A ferromagnesian mineral is olivine.
Silica	Silica refers to a chemical combination of silicon and oxygen.
Polymorphs	Polymorphs refer to two or more minerals having the same chemical composition but different crystalline structures. Exemplified by the diamond and graphite forms of carbon.
Quartzite	Quartzite refers to an extremely durable, nonfoliated metamorphic rock derived from pure sandstone and consisting primarily of quartz.
Schist	A coarse-grained, strongly foliated metamorphic rock that develops from phyllite and splits easily into flat, parallel slabs is schist.
Fine-grained rock	Fine-grained rock refers to a rock in which most of the mineral grains are less than 1 millimeter across or less than 1/16 mm.
Porphyroblast	A large crystal surrounded by a much finer grained matrix of other minerals in a metamorphic rock is a porphyroblast.
Gneiss	Gneiss refers to a coarse-grained, foliated metamorphic rock marked by bands of light-colored minerals such as quartz and feldspar that alternate with bands of dark-colored minerals. This alternation develops through metamorphic differentiation.
Limonite	Limonite refers to a type of iron oxide that is yellowish-brown when powdered.
Cleavage	The tendency of certain minerals to break along distinct planes in their crystal structures where the bonds are weakest is called cleavage. Cleavage is tested by striking or hammering a mineral, and is classified by the number of surfaces it produces and the angles between adjacent surfaces.
Deformation	General term for the processes of folding, faulting, shearing, compression, or extension of rocks as the result of various natural forces is called deformation.
Groundmass	The matrix of smaller crystals within an igneous rock that has porphyritic texture is referred to as groundmass.
Inclusion	A piece of one rock unit contained within another is called inclusion. Inclusions are used in relative dating. The rock mass adjacent to the one containing the inclusion must have been there first in order to provide the fragment.
Luster	Luster refers to the reflection of light on a given mineral's surface, classified by intensity and quality. The appearance of a given mineral as characterized by the intensity and quality with which it reflects light.
Muscovite	Transparent or white mica that lacks iron and magnesium is referred to as muscovite.
Phyllite	A foliated metamorphic rock that develops from slate and is marked by a silky sheen and medium grain size is phyllite.
Biotite	Biotite refers to iron/magnesium-bearing mica.
Hornfels	A hard, dark-colored, dense metamorphic rock that forms from the intrusion of magma into shale or basalt is called hornfels.

Go to **Cram101.com** for the Practice Tests for this Chapter.

Oxidation	Oxidation refers to the loss of electrons from a substance involved in a redox reaction; always accompanies reduction.
Iron	Iron is essential to all organisms, except for a few bacteria. It is mostly stably incorporated in the inside of metalloproteins, because in exposed or in free form it causes production of free radicals that are generally toxic to cells.
Atoll	An atoll is a type of low, coral island found in tropical oceans and consisting of a coral-algal reef surrounding a central depression. The depression may be part of the emergent island, but more typically is a part of the sea (that is, a lagoon), or very rarely is an enclosed body of fresh, brackish, or highly saline water.
Core	The innermost layer of the Earth, consisting primarily of pure metals such as iron and nickel is the core. The core is the densest layer of the Earth, and is divided into the outer core, which is believed to be liquid, and the inner core, which is believed to be solid.
Outcrop	Any place where bedrock is visible on the surface of the Earth is referred to as outcrop.
Metamorphic grade	A measure used to identify the degree to which a metamorphic rock has changed from its parent rock is a metamorphic grade. A metamorphic grade provides some indication of the circumstances under which the metamorphism took place.
Staurolite	Staurolite is a red brown to black mostly opaque nesosilicate mineral with a white streak. It crystallizes in the monoclinic crystal system, has a Mohs hardness of 7 to 7.5 and a rather complex chemical formula. Iron, magnesium and zinc occur in variable ratios.
Carbonate	One of several minerals containing one central carbon atom with strong covalent bonds to three oxygen atoms and typically having ionic bonds to one or more positive ions is carbonate.
Mafic	Mafic refers to a term used to describe the amount of dark-colored iron and magnesium minerals in an igneous rock.
Index mineral	Index mineral refers to a mineral that is a good indicator of the metamorphic environment in which it formed. Used to distinguish different zones of regional metamorphism.
Sillimanite	Sillimanite is one of three alumino-silicate polymorphs, the other two being andalusite and kyanite. A common variety of sillimanite is known as fibrolite, so named because the mineral appears like a bunch of fibres twisted together when viewed under thin section or even by the naked eye.
Potassium	Potassium is a chemical element in the periodic table. It has the symbol K (L. kalium) and atomic number 19. Potassium is a soft silvery-white metallic alkali metal that occurs naturally bound to other elements in seawater and many minerals.
Kyanite	Kyanite is a typically blue silicate mineral, commonly found in aluminium-rich metamorphic pegmatites and/or sedimentary rock. Kyanite is a diagnostic mineral of the Blueschist Facies of metamorphic rocks.
Slate	A fine-grained, foliated metamorphic rock that develops from shale and tends to break into thin, flat sheets is called slate.
Aureole	Zone of contact metamorphism adjacent to a pluton is called the aureole.
Contact metamorphism	Contact metamorphism is the name given to the changes that take place when magma is injected into the surrounding solid rock (country rock). The changes that occur are greatest wherever the magma comes into contact with the rock because the temperatures are highest at this boundary and decrease with distance from it.
Metasomatism	Metasomatism refers to chemical alteration of a rock by the action of hydrothermal fluids.

Go to **Cram101.com** for the Practice Tests for this Chapter.

Calcareous	Calcareous refers to a sediment, sedimentary rock, or soil type which is formed from or contains a high proportion of calcium carbonate in the form of calcite or aragonite. It is also used to refer to relatively alkaline soil.
Lead	Lead is a chemical element in the periodic table that has the symbol Pb and atomic number 82. A soft, heavy, toxic and malleable poor metal, lead is bluish white when freshly cut but tarnishes to dull gray when exposed to air. Lead is used in building construction, lead-acid batteries, bullets and shot, and is part of solder, pewter, and fusible alloys.
Claystone	A rock made up exclusively of clay-sized particles is referred to as claystone.
Recrystalliz-tion	Recrystallization refers to the diagenetic process by which unstable minerals in buried sediment are transformed into stable ones.
Nonfoliated	Metamorphic rocks that do not exhibit foliation are nonfoliated.
Schistose	The texture of a rock in which visible platy or needle-shaped minerals have grown essentially parallel to each other under the influence of directed pressure is schistose.
Matrix	Matrix refers to the solid matter in which a fossil or crystal is embedded.
Strain	Strain refers to the change in the shape or volume of a rock that results from stress.
Organic compound	An organic compound is any member of a large class of chemical compounds whose molecules contain carbon, with the exception of carbides, carbonates, carbon oxides and gases containing carbon.
Hornblende	Hornblende refers to common amphibole frequently found in igneous and metamorphic rocks.
Foliation	The layering within metamorphic rocks is called foliation and it occurs when a strong compressive force is applied from one direction to a recrystallizing rock. This causes the platy or elongated crystals of minerals, such as mica and chlorite, to grow with their long axes perpendicular to the direction of the force.
Compound	An electrically neutral substance that consists of two or more elements combined in specific, constant proportions is a compound. A compound typically has physical characteristics different from those of its constituent elements.
Vein	A sheetlike deposit of minerals precipitated in fractures or joints that are foreign to the host rock is called a vein.
Fracture	Fracture refers to a crack or break in a rock. To break in random places instead of cleaving.
Fault	A fracture dividing a rock into two sections that have visibly moved relative to each other is a fault.
Migmatite	Migmatite refers to a rock that incorporates both metamorphic and igneous materials.
Metamorphic differentiation	The process by which minerals from a chemically uniform rock separate from each other during metamorphism and form individual layers within a new metamorphic rock is referred to as metamorphic differentiation.
Differentiation	Differentiation refers to separation of different ingredients from an originally homogeneous mixture.
Stage	Stage refers to the height of floodwaters in feet or meters above an established datum plane.
Glassy	A term used to describe the texture of certain igneous rocks, such as obsidian, that contain no crystals is referred to as glassy.
Lineation	A rock texture formed by alignment of rod-like features or grains is referred to as lineation.

Go to **Cram101.com** for the Practice Tests for this Chapter.

Ultramafic rock	Rock composed entirely or almost entirely of ferromagnesian minerals is referred to as ultramafic rock.
Ultramafic	Ultramafic refers to igneous rocks made mostly of the mafic minerals hypersthene, augite, and/or olivine.
Gabbro	Gabbro refers to any of a group of dark, dense, phaneritic, intrusive rocks that are the plutonic equivalent to basalt.
Correlation	The process of determining that two or more geographically distant rocks or rock strata originated in the same time period is referred to as correlation.
Marble	A coarse-grained, nonfoliated metamorphic rock derived from limestone or dolostone is referred to as marble.
Igneous rocks	Rocks that crystallize from molten material at the surface of the earth or within the earth are called igneous rocks.
Granite	A pink-colored, felsic, plutonic rock that contains potassium and usually sodium feldspars, and has quartz content of about 10% is granite. Granite is commonly found on continents but virtually absent from the ocean basins.
Facies	Facies refers to a portion of a rock unit that possesses a distinctive set of characteristics that distinguishes it from other parts of the same unit.
Vapor	Vapor refers to water in the gaseous state.
Fracture zone	Linear zone of irregular topography on the deep-ocean floor that follows transform faults and their inactive extensions is called the fracture zone.
Fold	Fold refers to a bend that develops in an initially horizontal layer of rock, usually caused by plastic deformation. Folds occur most frequently in sedimentary rocks.
Hydrothermal metamorphism	Hydrothermal metamorphism refers to chemical alterations that occur as hot, ion-rich water circulates through fractures in rock.
Solution	Usually slow but effective process of weathering and erosion in which rocks are dissolved by water is a solution.
Breccia	A clastic rock composed of particles more than 2 millimeters in diameter and marked by the angularity of its component grains and rock fragments is called breccia.
Fracturing	Cracking or rupturing of a body under stress is referred to as fracturing.
Strike	The direction or trend of a bedding plane or fault, as it intersects the horizontal is referred to as a strike.
Dip	The angle formed by the inclined plane of a geological structure and the horizontal plane of the Earth's surface is referred to as a dip.
Zoning	Orderly variation in the chemical composition within a single crystal is referred to as zoning.
Period	Period refers to a basic unit of the geologic time scale that is a subdivision of an era. Periods may be divided into smaller units called epochs.
Butte	Butte refers to a small, conspicuous, isolated hill bounded by cliffs.
Silicification	Silicification refers to a process whereby silica replaces the original material of a substance.
Intermediate	Intermediate refers to a descriptive term applied to igneous rocks that are transitional between basic and acidic with silica between 54% and 65%.

Ore	A mineral deposit that can be mined for a profit is called ore.
Ore mineral	The part of an ore, usually metallic, that is economically desirable is called ore mineral.
Detritus	Detritus is organic waste material from decomposing dead plants or animals.
Shear	Shear refers to stress that causes two adjacent parts of a body to slide past one another.
Foliated rock	Foliated rock refers to a metamorphic rock that displays foliation. Foliated rocks include slate, phyllite, schist, and gneiss.
Conglomerate	Conglomerate refers to a clastic rock composed of particles more than 2 millimeters in diameter and marked by the roundness of its component grains and rock fragments.
Clast	An individual grain or constituent of a rock is a clast.
Dike	Dike refers to a discordant pluton that is substantially wider than it is thick. Dikes are often steeply inclined or nearly vertical.
Progressive metamorphism	Progressive metamorphism refers to metamorphism in which progressively greater pressure and temperature act on a rock type with increasing depth in Earth's crust.
Schistosity	Schistosity refers to a type of foliation characteristic of coarser-grained metamorphic rocks. Such rocks have a parallel arrangement of platy minerals such as the micas.
Ore deposit	The same as a mineral reserve except that it refers only to a metal-bearing deposit is referred to as ore deposit.
Greenschist	A schist containing chlorite and epidote and formed by low-pressure, low-temperature metamorphism of mafic volcanic rocks is greenschist.
Batholith	Batholith refers to a massive discordant pluton with a surface area greater than 100 square kilometers, typically having a depth of about 30 kilometers. Batholiths are generally found in elongated mountain ranges after the country rock above them has eroded.
Geology	The scientific study of the Earth, its origins and evolution, the materials that make it up, and the processes that act on it is called geology.
River	A major branch of a stream system is referred to as a river.

Rock	Rock refers to a naturally formed aggregate of usually inorganic materials from within the Earth.
Facies	Facies refers to a portion of a rock unit that possesses a distinctive set of characteristics that distinguishes it from other parts of the same unit.
Fossil	A preserved remnant or impression of an organism that lived in the past is referred to as fossil.
Collection	The accumulation of precipitation into surface and underground areas, including lakes, rivers, and aquifers is a collection.
Fault	A fracture dividing a rock into two sections that have visibly moved relative to each other is a fault.
Foliated rock	Foliated rock refers to a metamorphic rock that displays foliation. Foliated rocks include slate, phyllite, schist, and gneiss.
Foliation	The layering within metamorphic rocks is called foliation and it occurs when a strong compressive force is applied from one direction to a recrystallizing rock. This causes the platy or elongated crystals of minerals, such as mica and chlorite, to grow with their long axes perpendicular to the direction of the force.
Strike	The direction or trend of a bedding plane or fault, as it intersects the horizontal is referred to as a strike.
Rocks	Aggregates of minerals or rock fragments are called rocks.
Schist	A coarse-grained, strongly foliated metamorphic rock that develops from phyllite and splits easily into flat, parallel slabs is schist.
Mountain	A large mass of rock projecting above surrounding terrain is called a mountain.
Terrace	A flat, steplike surface that lines a stream above the floodplain, often paired one on each side of the stream, marking a former floodplain that existed at a higher level before regional uplift or an increase in discharge caused the stream to erode into it is called the terrace.
Gravel	Rounded particles coarser than 2 mm in diameter are called gravel.
Loess	A load of silt that is produced by the erosion of outwash and transported by wind is loess. Much loess found in the Mississippi Valley, China, and Europe is believed to have been deposited during the Pleistocene Epoch.
Bench	The unstable, newly-formed front of a lava delta is a bench.
Limestone	Limestone is a sedimentary rock composed largely of the mineral calcite (calcium carbonate: $CaCO_3$). Limestone often contains variable amounts of silica in the form of chert or flint, as well as varying amounts of clay, silt and sand as disseminations, nodules, or layers within the rock.
Clast	An individual grain or constituent of a rock is a clast.
Matrix	Matrix refers to the solid matter in which a fossil or crystal is embedded.
Ash	Fine particles of pulverized rock blown from an explosion vent are called ash. Measuring less than 1/10 inch in diameter, ash may be either solid or molten when first erupted.
Cretaceous	The Cretaceous period is one of the major divisions of the geologic timescale, reaching from the end of the Jurassic period, about 146 million years ago (Ma), to the beginning of the Paleocene epoch of the Tertiary period (65.5 Ma).
Formation	A body of rock identified by lithic characteristics and stratigraphic position and is

Go to **Cram101.com** for the Practice Tests for this Chapter.

mappable at the earth's surface or traceable in the subsurface is a formation.

Alluvial   Alluvial refers to pertaining to material or processes associated with transportation and or subaerial deposition by concentrated running water.

Eolian   Eolian refers to of, produced by, or carried by wind.

Scale   The relationship between distance on a map and the distance on the terrain being represented by that map is a scale.

Till   Dominantly unsorted and unstratified drift, generally unconsolidated deposited directly by and underneath a glacier without subsequent reworking by meltwater, and consisting of a hetergeneous mixture of clay, silt, sand, gravel, stones, and boulders is called till.

Soil   Soil refers to the top few meters of regolith, generally including some organic matter derived from plants.

Landform   Landform refers to any physical, recognizable form or feature on the earth's surface, having a characteristic shape and range in composition, and produced by natural causes.

Dip   The angle formed by the inclined plane of a geological structure and the horizontal plane of the Earth's surface is referred to as a dip.

Carbon   Carbon is a chemical element in the periodic table that has the symbol C and atomic number 6. An abundant nonmetallic, tetravalent element, carbon has several allotropic forms.

Horizon   Horizon refers to levels within a soil profile that differ structurally and chemically. Generally divided into A, B, C, E, and 0 horizons.

Fracture   Fracture refers to a crack or break in a rock. To break in random places instead of cleaving.

Pluton   An intrusive rock, as distinguished from the preexisting country rock that surrounds it is called pluton.

Dike   Dike refers to a discordant pluton that is substantially wider than it is thick. Dikes are often steeply inclined or nearly vertical.

Element   A chemical element, often called simply element, is a chemical substance that cannot be divided or changed into other chemical substances by any ordinary chemical technique. An element is a class of substances that contain the same number of protons in all its atoms.

Biostratigraphy   Biostratigraphy refers to the study and classification of rocks and their history based on their fossil content.

Stratigraphic   The study of rock strata, especially of their distribution, deposition, and age is called stratigraphic.

Correlation   The process of determining that two or more geographically distant rocks or rock strata originated in the same time period is referred to as correlation.

Zonation   The distribution of organisms in bands or regions corresponding to changes in ecological conditions along a continuum, for example, intertidal zonation and elevational zonation.

Ecology   Ecology is the scientific study of the distribution and abundance of living organisms and how these properties are affected by interactions between the organisms and their environment.

Numerical age   Age given in years or some other unit of time is the numerical age.

Metamorphic   Metamorphic refers to the term from the Greek 'meta' and 'morph', commonly occurs to rocks which are subjected to increased heat and/or pressure. Also applies to the conversion of snow into glacial ice.

Deformation   General term for the processes of folding, faulting, shearing, compression, or extension of

Go to **Cram101.com** for the Practice Tests for this Chapter.

Go to **Cram101.com** for the Practice Tests for this Chapter.
And, **NEVER** highlight a book again!

	rocks as the result of various natural forces is called deformation.
Igneous	Igneous rocks are formed when molten rock (magma) cools and solidifies, with or without crystallization, either below the surface as intrusive (plutonic) rocks or on the surface as extrusive (volcanic) rocks. This magma can be derived from either the Earth's mantle or pre-existing rocks made molten by extreme temperature and pressure changes.
Strain	Strain refers to the change in the shape or volume of a rock that results from stress.
Stage	Stage refers to the height of floodwaters in feet or meters above an established datum plane.
Ore deposit	The same as a mineral reserve except that it refers only to a metal-bearing deposit is referred to as ore deposit.
Ore	A mineral deposit that can be mined for a profit is called ore.
Mineral	A naturally occurring, usually inorganic, solid consisting of either a single element or a compound, and having a definite chemical composition and a systematic internal arrangement of atoms is referred to as a mineral.
Stratification	The arrangement of sedimentary rocks is called stratification.
Homogeneous	Homogeneous refers both to animals and plants, of having a resemblance in structure, due to descent from a common progenitor with subsequent modification.
Geology	The scientific study of the Earth, its origins and evolution, the materials that make it up, and the processes that act on it is called geology.
Bedding	The division of sediment or sedimentary rock into parallel layers that can be distinguished from each other by such features as chemical composition and grain size is bedding.
Stratum	A stratum is a layer of rock or soil with internally consistent characteristics that distinguishes it from contiguous layers. Each layer is generally one of a number of parallel layers that lie one upon another, laid down by natural forces.
Strata	Parallel layers of sedimentary rock are called strata.
Lead	Lead is a chemical element in the periodic table that has the symbol Pb and atomic number 82. A soft, heavy, toxic and malleable poor metal, lead is bluish white when freshly cut but tarnishes to dull gray when exposed to air. Lead is used in building construction, lead-acid batteries, bullets and shot, and is part of solder, pewter, and fusible alloys.
Column	A feature found in caves that is formed when a stalactite and stalagmite join is referred to as a column.
Plate	Plate refers to rigid parts of the Earth's crust and part of the Earth's upper mantle that moves and adjoins each other along zones of seismic activity.
Rank	Rank refers to a coal's carbon content depending upon its degree of metamorphism.

Go to **Cram101.com** for the Practice Tests for this Chapter.